LIPID ANALYSIS

The INTRODUCTION TO BIOTECHNIQUES series

Editors:

J.M. Graham Merseyside Innovation Centre, 131 Mount Pleasant, Liverpool L3 5TF

D. Billington School of Biomolecular Sciences, Liverpool John Moores University, Byrom Street, Liverpool L3 3AF

Series adviser:

P.M. Gilmartin Centre for Plant Biochemistry and Biotechnology, University of Leeds, Leeds LS2 9JT

CENTRIFUGATION
RADIOISOTOPES
LIGHT MICROSCOPY
ANIMAL CELL CULTURE
GEL ELECTROPHORESIS: PROTEINS
PCR
MICROBIAL CULTURE
ANTIBODY TECHNOLOGY
GENE TECHNOLOGY
LIPID ANALYSIS

Forthcoming titles

GEL ELECTROPHORESIS: NUCLEIC ACIDS
LIGHT SPECTROSCOPY
MEMBRANE ANALYSIS
PLANT CELL CULTURE

LIPID ANALYSIS

F.W. Hemming and J.N. Hawthorne
Department of Biochemistry, Medical School, Queen's Medical
Centre, Nottingham NG7 2UH, UK

Taylor & Francis
Taylor & Francis Group

LONDON AND NEW YORK

© Taylor & Francis Publishers Limited, 1996

First published 1996

A CIP catalogue record for this book is available from the British Library.

ISBN 1 872748 98 8

Published by Taylor & Francis
2 Park Square, Milton Park, Abingdon, Oxon, OX14 4RN
270 Madison Ave, New York NY 10016

Transferred to Digital Printing 2008

Typeset by Chandos Electronic Publishing, Stanton Harcourt, UK.

Publisher's Note
The publisher has gone to great lengths to ensure the quality of this reprint but points out that some imperfections in the original may be apparent

Contents

Abbreviations   x
Preface   xi

1. Introduction   1

2. Basic Techniques   5

Extraction methods   5
  Lipid binding in tissues   5
  Solvents for lipid extraction   5
  Some precautions   6
  Storage of lipid extracts   6
  Saponification   6
Chromatographic procedures   7
  General considerations   7
  Partition chromatography   12
  Adsorption chromatography   29
  Ion exchange chromatography   30
  Complexing chromatography   32
Spectrometric methods   32
  UV–visible spectrometry   33
  IR spectroscopy   35
  NMR spectroscopy   35
  Mass spectrometry   36
Radioisotopic methods   37
  Choice of radioisotope   37
  Autoradiography   39
  Scintillation counting   41
  Gas ionization detection   45
  Gamma counting   46
  Radioisotopes in lipid studies   47
  RIA and related techniques   49
Immunochemical methods   51
  Antibodies to lipids   51
  Hemagglutination in detection and assay   52
  Immunostaining and TLC   54
  Immunoradiometric assays   55
  Enzyme-linked immunosorbent assays   56

Analytical methods involving enzymes   57
  Phospholipases   57
  Stereospecific analysis of triacylglycerol   59
  Enzymes in quantifying lipids   60
Presentation of lipids in aqueous systems   60
Care and safety during the analysis of lipids in biological
    materials   61
  General considerations   61
  Chemical hazards   62
  Fire hazards   65
  Radioisotope hazards   66
  Biological hazards   67
  Other hazards   68
  Personal preparedness and protection   69
References   70

3. Hydrocarbons   73

Biological significance   73
Structures   73
Detection   74
Isolation and purification   75
Quantitation   78
References   78

4. Alcohols, Phenols, Aldehydes, Ketones and Quinones   79

Alcohols and phenols   79
  General and biological significance   79
  Structures   80
  Detection   86
  Isolation and purification   89
  Quantitation   91
Aldehydes and ketones   92
  General and biological significance   92
  Structures   93
  Detection   93
  Isolation and purification   94
  Quantitation   95
Quinones   95
  General and biological significance   95
  Structures   96
  Detection   96
  Isolation and purification   99
  Quantitation   100
References   100

5. Fatty Acids and Prostaglandins   **101**

Fatty acids   101
  Types of fatty acids found naturally   101
  Separation methods for fatty acids   101
  GLC of fatty acids   103
  HPLC of fatty acids   105
Prostaglandins and leukotrienes (eicosanoids)   105
  Analysis of eicosanoids   106
References   106

6. Esters   **107**

Acylglycerols   107
  Separation of acylglycerols   107
  Analysis of acylglycerols   108
  Fatty acids of acylglycerols   108
Wax esters   110
Alkylglycerols   110
  Analytical methods for alkylglycerols   111
References   111

7. Phospholipids, Sulfolipids and Related Compounds   **113**

Phospholipids with amino-groups   113
  Hydrolytic procedures in analysis   114
  Estimation of phosphate   115
  Separation of phospholipids   115
Phosphoinositides   116
  Phosphoinositide and inositol phosphate metabolism   118
  Separation of the phosphoinositides   119
  Receptor-linked phosphoinositide changes   119
  Separation and estimation of inositol phosphates   120
  Estimation of diacylglycerol   122
Phosphatidylinositol glycans   122
  Detection of PI glycans   123
Phosphonolipids   124
  Separation and analysis of phosphonolipids   124
Alkylether phospholipids   125
  Analysis of alkylether phospholipids   125
Sulfolipids   126
  Lipid sulfates   126
  Sulfonolipids   126
  Analytical methods   127
References   127

8.   Glycolipids   129

Glycosphingolipids   129
  Biological significance   129
  Structures   130
  Detection   132
  Isolation and purification   134
  Quantitation   139
Polyisoprenyl phosphosugars   142
  Biological significance   142
  Structures   142
  Detection   143
  Isolation and purification   145
  Quantitation   145
Glycosylglycerides and related compounds   146
  Biological significance   146
  Structures   147
  Detection   148
  Isolation and purification   148
  Quantitation   149
Phosphoglycolipids, lipoteichoic acids and related
  compounds   149
  Biological significance   149
  Structures   150
  Detection, isolation, preparation and quantitation   150
References   151

9.   Lipoproteins   153

Plasma lipoproteins   153
  Structures and biological significance   153
  Detection   157
  Isolation and purification   158
  Quantitation   160
Fatty acyl proteins and prenyl proteins   161
  Biological significance   161
  Structures   161
  Detection   163
  Isolation and purification   163
  Quantitation   164
References   164

Appendices

Appendix A: Glossary   165
Appendix B: Suppliers   167
Appendix C: Further reading   171

Index   173

Abbreviations

ACDP	Advisory Committee on Dangerous Pathogens
COSHH	Control of Substances Hazardous to Health
CM	chylomicrons
DEAE	diethylaminoethyl
ECD	electron capture detector
ELISA	enzyme-linked immunosorbent assay
FID	flame ionization detector
GC	gas chromatography
GC–MS	gas chromatography–mass spectrometry
GLC	gas–liquid chromatography
GM	Geiger–Müller
HDL	high density lipoproteins
HETP	height equivalent to a theoretical plate
HIV	human immunodeficiency virus
HMIP	HM Inspectorate of Pollution
HPAEC	high-performance anionic exchange chromatography
HPLC	high-performance liquid chromatography
i.d.	internal diameter
IDL	intermediate density lipoproteins
IR	infrared
IRMA	immunoradiometric assay
LC	liquid chromatography
LDL	low density lipoproteins
LSC	liquid scintillation counting
MAb	monoclonal antibodies
MS	mass spectometry
NMR	nuclear magnetic resonance
PBS	phosphate-buffered saline
PG	prostaglandin
PI	phosphatidylinositol
PM	photomultiplier
POPOP	1,4-bis-2-(5-phenyloxazolyl)-benzene
PPO	2,5-diphenyloxazole
RIA	radioimmunoassay
RPO	Radiation Protection Officers
SDS–PAGE	sodium dodecyl sulfate–polyacrylamide gel electrophoresis
SI	Système International
TLC	thin-layer chromatography
VLDL	very low density lipoproteins

Preface

In following our research interests in lipid biochemistry, we have been impressed not only by the wide range of lipid-soluble compounds found in living systems but also by the great variety of biological phenomena in which lipid-soluble compounds play an important role.

At its simplest, this is seen in the passive, protective hydrophobic barrier of waxes found in many insects and plant leaves or in the energy store of triacylglycerols found in many eukaryotic cells. All biological membranes are composed primarily of phospholipids which may fulfill several functions. They provide a permeability barrier, so important in compartmentation of eukaryotic cells, but also provide a lipid environment within which many proteins involved in cell- (and organelle-) surface phenomena function. The exciting discovery of the second messenger activity of inositol phosphates and of diacylglycerol drew attention to the critical part played by phosphatidylinositol in the transduction of messages across the plasma membrane of animal cells. A different activity is shown by the glycolipids of the plasma membrane in the form of a key function in cell recognition phenomena important, for example, in embryonic development and in malignancy.

Inside the cell, several lipids are anchored in membranes by polyisoprenoid chains in a manner appropriate to their role in electron transport (e.g. the quinones) or in protein glycosylation (the dolichols). However, not all biologically active lipids are membrane components and some, such as steroid hormones, prostaglandins and pheromones, can in trace quantities stimulate large biochemical and physiological changes.

This book deals with the basic aspects of the analytical techniques needed to investigate the complexity of lipids in living cells. On the one hand, it is intended to provide sufficient information to understand the experimental basis of and to assess observations already published. On the other hand, it aims to equip the lipid researcher with a background to techniques appropriate for any experimental studies planned.

We have deliberately not provided detailed protocols. These can be ephemeral and are probably best prepared by the individual based on the primary scientific literature or books on specific methodologies and bearing in mind local circumstances.

It is anticipated that the book will be of most use to advanced undergraduates or junior research workers setting out to get to grips with the analysis of a lipid. It is hoped that the arrangement of chapters will be 'user friendly', particularly for those with special interests in the analysis of one specific group of lipids or in the applications of one particular technique.

F.W. Hemming
J.N. Hawthorne

1 Introduction

Recognition of the diversity and biological importance of lipids or of lipid modifications of other compounds (e.g. of proteins) continues to increase. In addition, variations in concentration of some lipids of animal tissues have important consequences for the health of the animal. This book aims to explain and discuss the principles of the methods used currently in the analysis of lipids to assist those wishing to understand or plan experiments in this area.

The term lipid is used broadly to describe compounds of biological origin that will partition into an organic solvent that is immiscible with water. Analysis is assumed to involve recognition, isolation and quantitative assay of a compound or group of compounds. Much of this book, therefore, consists of a discussion of the most suitable methods used in elucidating these aspects of analysis for each of the major groups of lipids.

Lipid-soluble compounds are grouped on a chemical basis. This has led to chapters on hydrocarbons, alcohols (plus phenols, quinols, quinones and aldehydes), fatty acids, esters, phospholipids and glycolipids. The final chapter deals with lipoproteins which contain several of these groups of compounds and present special analytical challenges. Since the analysis of several of these compounds involves the application of common techniques, it has proved convenient to precede the compound-based chapters with one dealing in general terms with the techniques used in lipid analysis. The successful analyst will ensure not only that he/she thoroughly understands the basis, and strengths and weaknesses of the methodology described in Chapter 2, but also is well versed with regard to the manufacturer's instructions for any equipment that is to be used.

Armed with this information, it is wise to prepare an analytical protocol and to attempt to anticipate particular needs or potential problems that the protocol may present. This book deliberately avoids providing protocols but recommends that these be written by the analyst with the aid of a detailed techniques text or original paper describing the analysis and taking account of the local situation.

Chapters 3–9 summarize much of the general information pertinent to preparing an appropriate protocol for individual compounds or groups of compounds.

Each of Chapters 3–9 opens with a brief account of the biological significance of the group of lipids under consideration. This biological context helps in assessment of the types of analytical questions that may need to be answered.

Analytical methods involve the application of physical and chemical techniques to complex mixtures. Clearly, the analyst will be best able to develop new methods, to optimize existing methods or simply to apply protocols derived from others if he/she has at least a basic appreciation of the chemical structures and properties of the compounds to be analyzed. For this reason, each of Chapters 3–9 includes a section dealing with this aspect. In describing the structures, internationally agreed nomenclature based on the recommendations of the International Union of Pure and Applied Chemistry (IUPAC) and of the International Union of Biochemistry (IUB) has been used.

This section is followed by essentially qualitative aspects of analysis (detection, separation, purification) before discussing quantitation of particular lipids. This is not to relegate quantitative data but simply recognizes that some aspects of qualitative analysis may be essential features of quantitation. In fact, it is important to appreciate that most analyses should be aimed at providing good quantitative data. In order to be widely interpretable, this data should be expressed in SI units (recommended by the Système International d'Unités) with sufficient information to allow judgement of the reliability and repro-ducibility of the results. To this end, the protocol used in the determination should be sufficiently clear to allow someone else to repeat the work. Random experimental errors should be assessed and reported by determining the mean and standard deviation of a reasonable number of replicable determinations. Systematic errors cannot be treated in this statistical way. However, attempts should be made to assess them, for example, by adding known amounts of internal standards during the analysis and setting up appropriate control experiments. If systematic errors cannot be eliminated they should be quantitated and taken into account. For example, if an internal standard is routinely being determined at 75% of the quantity added, it is reasonable to report the content of that analyte as 133% (100/75) of that actually measured.

The analyst will also need to be aware of the sensitivity and specificity of an assay. It is reassuring to learn that analytical data reported are well within the sensitivity of the method used and that they are unlikely to be compromised by the presence of other compounds. In attempting to maximize one of these aspects, it is possible that sacrifices will have to be made with regard to the other.

All quantitative analytical methods should be supported by good calibration data, preferably calibration curves covering the range of analyte concentrations being reported. Having taken this and other precautions emphasized above, it is incumbent upon the analyst to judge realistically the precision of the determination being reported. Stating data to the fifth figure (e.g. 15.473 mg g^{-1} should be avoided if the assay has a precision of $\pm 1\%$. A more realistic statement would be 15.5 mg g^{-1}.

Finally, when reporting the results of lipid analysis, presentation of data and conclusions are all important. Currently available software for computers and wordprocessors assists greatly in this. Graphs, pie charts and histograms often get the essential message of comparative assays across to others more readily than do tables of numbers, especially if used as a visual aid to an oral report. Flow sheets may be a more accessible presentation of experimental protocol than text. It is essential for the analyst to consider the potential audience or readership of his/her analytical report and select an appropriate form of presentation.

2 Basic Techniques

2.1 Extraction methods

2.1.1 Lipid binding in tissues

Nonpolar lipids such as triacylglycerols are bound in tissues to other lipids or to the hydrophobic regions of proteins by relatively weak Van der Waals or hydrophobic bonds. More polar lipids such as phospholipids may be bound to proteins by hydrogen bonds and electrostatic association, as well as by hydrophobic interactions. Very rarely, lipids are covalently bound to polysaccharides or proteins. Examples are the lipopolysaccharides of bacterial cell walls and the phosphatidylinositol glycans in which specific proteins are anchored to the plasma membrane via the phospholipid.

2.1.2 Solvents for lipid extraction

Folch showed that a mixture of a nonpolar and a polar solvent was most effective in extracting lipids from tissues. The Bligh and Dyer modification of his method (see ref. 1) uses a chloroform–methanol–water mixture (1:2:0.8 by vol.). About 6 ml of this solvent mixture is used per gram of tissue for homogenization at room temperature. After centrifugation to remove the tissue residue, a further one volume each of chloroform and water are added to the supernatant to form a two-phase system. Water-soluble contaminants are thus removed in the upper aqueous layer while lipids are concentrated in the lower organic phase. A second extraction of the tissue residue with chloroform–methanol–0.2 M HCl (1:2:0.8 by vol.) then extracts the more tightly bound phosphoinositides. Washing is done by adding one volume of chloroform and one of 2 M KCl. The KCl is required to prevent loss of polyphosphoinositides in the aqueous phase.

5

2.1.3 Some precautions

Many lipid molecules contain double bonds, for instance the polyunsaturated fatty acids, and these are subject to peroxidation. Some workers therefore bubble nitrogen through the extracting solvents before use and try to perform their procedures under nitrogen. An anti-oxidant such as butylated hydroxytoluene may also be added to the solvents at a concentration of around 0.05% (w/v).

Acid or alkaline conditions can hydrolyze some lipids, so it is important to keep extracts at neutral pH, for example after using acidified chloroform–methanol. A second wash may be required, and can be performed conveniently with a 'synthetic upper phase', made by mixing the solvents in the proportions originally used, which would be chloroform (2 vol.), methanol (2 vol.), water (0.8 vol.) and 2 M KCl (1 vol.). The pH of the upper phase is checked. Neutral pH is particularly important if an extract is to be dried down by rotary evaporation. Lipids should not be left in the dry or nearly dry state for long, but redissolved in a suitable solvent (see Section 2.1.4).

Some plant tissues contain phospholipases which may remain active during extraction. Procedures to avoid this, including extraction with hot isopropanol, are described by Kates [1].

2.1.4 Storage of lipid extracts

Analysis of lipids is, of course, best performed on fresh material. If lipids must be stored, any moisture is first removed by azeotropic distillation with benzene. Addition of a little benzene followed by rotary evaporation will remove water and benzene together. The lipid is then dissolved in freshly distilled chloroform containing 25–50% methanol and stored at –20°C. Since chloroform can give rise to acidic breakdown products, particularly in sunlight, it should not be used for periods of storage longer than a few weeks. Longer-term storage is best done in benzene–methanol mixtures containing an anti-oxidant (Section 2.1.3), if possible at –40°C or lower temperatures.

2.1.5 Saponification

For analytical purposes it may be necessary to release all the fatty acids from a complex lipid mixture. Where only O-acyl esters are present this is readily done by alkaline hydrolysis or methanolysis. If plasmalogens are present the mixture is first heated in methanolic HCl to release the unsaturated ethers as dimethyl acetals. Further information is given in Section 7.1.1.

2.2 Chromatographic procedures

2.2.1 General considerations

Chromatography has featured prominently in lipid analysis for many years. In fact the term (literally 'colored writing' from the Greek) was first used in 1903 by the Russian botanist, Tswett, to describe the separation of lipid-soluble plant pigments on a column of calcium carbonate.

The characteristics of lipids that are relevant to their chromatographic separation include their degree of polarity, their degree of ionization and, in special situations, their ability to bind specifically and with quite high affinity to another compound.

The polarity of a lipid significantly affects its volatility, solubility and nonspecific binding to other polar materials. Polar molecules have an unbalanced distribution of electrons among their component atoms sufficient to produce a dipole. The resulting partial ionic nature of the lipid has several relevant consequences. Intermolecular binding of partial positive to partial negative charges reduces volatility. Similar interactions with polar solvents increase solubility and with solid polar materials increase retention by some chromatographic stationary phases.

Those lipids that are able to ionize fully to give a complete positive or complete negative charge take the concept of polarity further. They have very low volatility and if dissolved in a solvent that allows full ionization they are very polar. Polar solvents allow ionization especially if they contain acids or bases of opposite charge to that of the ionizing group or the lipid. Lists of lipids and solvents arranged in relation to their polarity are presented in *Tables 2.1* and *2.2* respectively.

The ability of a lipid to bind specifically and with high affinity to a particular compound forms the basis of affinity chromatography of lipids. The compound is immobilized on an appropriate particulate support to form the stationary phase. Immobilized antibodies and enzymes, for example, often feature in affinity chromatography.

The art of good chromatography is dependent primarily upon selection of the appropriate stationary and mobile solvent phases to take advantage of differences in polarity, ionization or specific affinity between the components of the mixture that are to be separated.

These selections generally fall into one of three types of chromatography, i.e. partition, adsorption or ion-exchange. These categories best describe the main features of the chromatographic system to be used, but rarely does one category operate without some contribution from the others.

TABLE 2.1: Some groups of lipids arranged in order of polarity

Least polar (most hydrophobic)

Hydrocarbons
Fatty esters (simple)
Aldehydes
Triacylglycerols
Fatty alcohols
Fatty acids
Quinones
Sterols
Diacylglycerols
Monoacylglycerols
Cerebrosides (monoglycosylceramides)
Glycosyldiacylglycerols
Sulfolipids
Phosphatidic acids
Phosphatidyl ethanolamine
Phosphatidyl serine
Phosphatidyl choline
Phosphatidyl inositol
Sphingomyelin

Most polar (most hydrophilic)

TABLE 2.2: Major solvents of lipids arranged in order of polarity

Least polar (most hydrophobic, least eluotropic)

Hexane
Cyclohexane
Carbon tetrachloride
Benzene
Diethyl ether
Chloroform (trichloromethane)
Dichloromethane
Chloroform (containing 1% ethanol as stabilizer)
Dioxane
Acetone
Acetonitrile
Pyridine
Butanols
Propanols
Ethanol
Methanol
Acetic acid (ethanoic acid)
Water

Most polar (most hydrophilic)

It is usually convenient to further classify chromatographic systems according to the special features of the physical arrangement that has been adopted. The stationary phase is usually held in position either as a packing of small particles in a column inside a cylindrical container or as a thin layer on a flat surface. Solvent then passes through the column or over the layer. This gives rise to the common terms column chromatography and thin-layer chromatography (TLC) (*Figure 2.1*). Paper chromatography is closely related to TLC, which in fact it predates, for the thin sheets of paper or materials bound to them act as the stationary phase. Column chromatography can also be further divided into liquid chromatography (LC) in which the mobile phase is a liquid and gas chromatography (GC) in which the mobile phase is a gas. In both cases the stationary phase may be a liquid bound to a solid support giving rise to partition chromatography or the solid itself giving rise to adsorption chromatography.

In column chromatography the separated lipids are washed (eluted) from the column one after the other. This may involve a continuous single solvent or a mixture of solvents (isocratic elution) or a gradually (or stepwise) changing mixture of solvents (gradient elution). For analytical work, aimed primarily at measuring the amount of each component in the lipid mixture, the most convenient arrangement (*Figure 2.2*) is to have attached to the outlet a flow-through device hooked up to other equipment which together are capable of detecting and measuring the amount or concentration of solute leaving the column. A wide range of sensitive devices (detectors) is used routinely which measure some change in the solution leaving the column which is characteristic of the presence of the solute(s) in question. Some detectors are sensitive to changes in light and in particular measure changes in (a) absorption of light at selected wavelengths, (b) fluorescence or (c) refractive index. Potentiometric, conductometric, coulometric and amperometric detectors are also available to measure electrochemical changes. Radioactive lipids may be assayed by flow-through radiodetectors. Alternatively if no suitable flow-through detector is available the solution (eluent) leaving the column is collected in a series of small fractions and these are assayed for the presence of solute(s) one fraction at a time. Automated fraction collectors which collect successive fractions of eluent on a time or volume or drop-count or detected peak basis remove the inconvenience of manual fraction collection.

Whichever method is used an elution profile is prepared showing the variation in the amount of solute(s) eluted against the volume of eluent passed or against the time elapsed, both since the chromatography began (*Figure 2.3*).

(a) Column chromatography

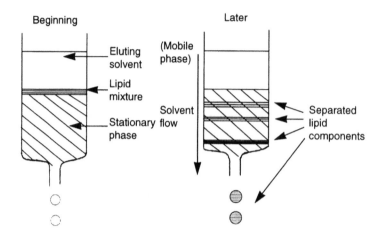

(b) Thin-layer chromatography

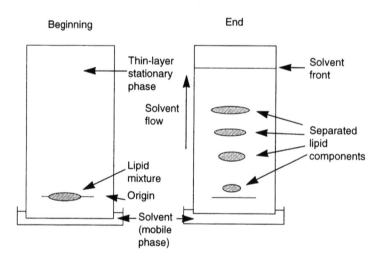

FIGURE 2.1:   Basic aspects of column (**a**) and thin-layer (**b**) chromatography.

In TLC the movement of the mobile phase across the layer of stationary phase is usually stopped before it reaches the end of the layer (*Figure 2.1*). After evaporation of the solvent, detection of the separated solutes normally depends on characteristics such as their color or their response to specific staining procedures. The intensity of color or stain may be assessed by scanning devices. However, TLC is often used for qualitative rather than quantitative assessment of separations as it is more difficult to achieve high accuracy and good

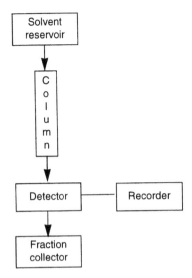

FIGURE 2.2:   *Basic arrangement of equipment for column chromatography.*

reproducibility with it than with solutions eluted from chromatographic columns.

In all chromatographic systems it is wise to protect the lipids and solvents from light and from oxidative or other chemical changes which might be catalyzed by surfaces of small particles. In particular, solvents such as diethyl ether should be rendered free of peroxides before use and esters should not be exposed to alkaline surfaces for long periods.

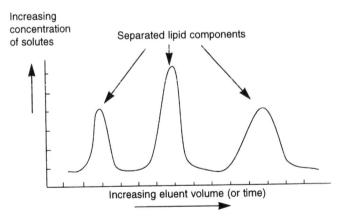

FIGURE 2.3:   *Elution profile of components of lipid mixture eluted separately from a chromatography column.*

2.2.2 Partition chromatography

In partition chromatography lipids are separated according to differences in their partition coefficients between two immiscible solvents. One of the solvents is bound to an inert solid support to create the stationary phase while the other flows over this to create the mobile phase. It is conventional to describe the arrangement in which the stationary phase is more polar than the mobile phase as normal phase partition chromatography. An arrangement in which the stationary phase is less polar than the mobile phase is termed reversed phase partition chromatography.

Several factors contribute to efficient separation, identification and quantification of solutes in partition chromatography. It is usually best if the mixture of solutes is applied in as small a volume as possible. Each solute partitions between the two solvents and because one solvent is continuously mobile, the solute(s) move(s) as a band(s) across the stationary phase in the direction of the mobile phase (*Figure 2.3*). During this process band broadening occurs due to diffusion effects. This may lead to overlap and hence poor separation of adjacent bands of solutes. It may also result in broad bands of low concentration of solutes, especially of slow moving solutes, which may be difficult to detect and measure.

Band broadening is accentuated primarily by problems of mass transfer and, in column chromatography, by eddy diffusion and excessive 'dead space' in the system. Mass transfer refers to the rate of diffusion of solute molecules between the stationary and mobile phases. Slow flow rates of mobile phase, thin layers of stationary phase and small, nonporous support particles (resulting in small spaces between particles and hence narrow channels through which the mobile phase can flow) tend to reduce mass transfer problems. Eddy diffusion can often be caused by a wide range of size of channels between particles, leading to variable flow of mobile phase and hence swirling of solvent when the channels join. This can be minimized by using evenly sized spherical support particles packed uniformly in the column. The effects of 'dead space' can be reduced by using 'zero dead volume' fittings and narrow bore connecting tubing between the column and post-column devices. Under normal conditions the simple diffusion of solute molecules in both phases is relatively slow and rarely contributes significantly to band broadening.

The position of the band (peak) on the chromatogram (*Figure 2.4*) is a useful characteristic of the solute. Providing all chromatographic parameters remain constant the position should be reproducible. In column chromatography it is measured usually in terms of retention

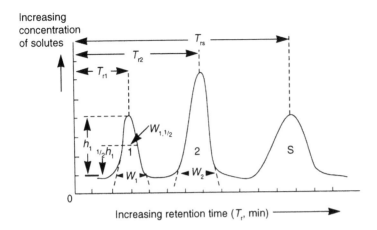

FIGURE 2.4: *Conventions used to describe column chromatographic performance in separating components 1, 2 and a standard (S).* T_{r1}, T_{r2} *and* T_{rs}, *retention times for peaks 1, 2 and S: h, height of peak 1; W_1, W_2, width of base of peaks 1 and 2; $W_{1, 1/2}$, width of peak 1 at half maximum height, $1/2\,h_1$.*

time (T_r), the time elapsing between application of the sample and the elution at highest concentration (i.e. at the top of the peak in *Figure 2.4*) of the component of the sample of interest. Alternatively the elution (retention) volume, E_v (R_v), the volume of eluting solvent required to elute the component of interest from the column, can be measured.

It may be difficult for technical reasons to hold all the parameters constant from one chromatography run to the next. For this reason it is often helpful to include in the applied sample a standard substance of known chromatographic mobility. In this situation the relative retention $(T_{r1}/T_{rs}$ or $V_{r1}/V_{rs})$ will be a more reliable characteristic than the absolute retention $(T_{r1}$ or $V_{r1})$. The identity of the retention time (volume) of an unknown solute relative to that of a known compound in several different chromatographic systems may be an important part of the characterization of the unknown.

The resolution of a mixture of two solutes into its components (*Figure 2.4*) may be described in terms of the resolution index (R_s). This is given by the relationship (with reference to *Figure 2.4*):

$$R_s = \frac{\text{twice the distance between the peaks}}{\text{sum of the base width of the peaks}}$$

or

$$R_s = \frac{2\,(T_{r2} - T_{r1})}{W_2 + W_1}.$$

Usually a value of 1.5 for R_s is considered an acceptable separation.

The efficiency of a chromatography column may be described in terms of the number, N, of theoretical plates to which the column is equivalent under defined conditions. The term 'theoretical plates' relates to the concept of a series of hypothetical layers of the two phases across which a solute is in equilibrium, to explain the shape of the peak achieved. The value of N is derived from:

$$N = 16\,\frac{T_r^2}{W}$$

Clearly the higher the value of N, the smaller the amount of peak broadening that occurs at a particular retention time and therefore the better the separation of two adjacent peaks. In fact it is often easier to measure the width of a peak at half its height ($W_{1/2}$) rather than at its base. The equation then becomes:

$$N = 5.54\,\frac{T_r^2}{W_{1/2}}$$

In general N increases with the following changes in parameters:

(i)   increase in the surface area of the stationary phase (i.e. decrease in support particle size);
(ii) decrease in flow rates;
(iii) decrease in viscosity of solvents;
(iv) increase in temperature;
(v)   decrease in size of solute molecule;
(vi) increase in column length.

When comparing columns it is usual to take column length, L, into account by referring to L/N, sometimes called the height equivalent to a theoretical plate (HETP). Efficient columns have small values of HETP, preferably in the range of 1–10. Given a good column with a small HETP the operator can usually optimize the separation of solutes within an acceptable time by attending to the composition of the mobile phase, flow rate and temperature.

Several developments of partition chromatography have led to different branches that depend on the principles already discussed but can be seen to be different technically. These branches are normal liquid-liquid chromatography (LC), high-performance (high-pressure) liquid chromatography (HPLC), gas-liquid chromatography (GLC) and thin-layer chromatography (TLC), and these are now discussed in turn.

Normal liquid–liquid chromatography (LC). This is the simplest and cheapest form of column chromatography. Columns are usually packed by hand. Solvent flow is often due to gravity alone, although with fine particle packings or long columns, a solvent pump may be needed and in any case this usually guarantees a uniform flow rate. This may be particularly important if gradients of mixtures of solvents of different viscosity are used or if addition of the lipid mixture initially partially blocks solvent flow. For normal phase partition LC the stationary phase is usually water bound to particles of cellulose and an immiscible less polar solvent is used as the mobile phase. However normal phase partition LC is rarely employed in lipid fractionations although it may be useful for the separation of some polar lipids.

On the other hand reversed phase partition LC has proved to be very useful for separation of most lipid classes. At its simplest the nonpolar stationary phase is usually held by hydrophobic forces on an inert support of silanized celite. Often the stationary phase is a hydrocarbon, for example heptane or liquid paraffin or a high-boiling-point silicone fluid. The mobile phase is commonly acetonitrile, propanol or acetone – each containing various proportions of methanol or water depending on the lipids to be separated. Sometimes these mobile phases require saturation with the stationary phase to avoid 'bleeding' of stationary phase into the eluent. Some use has been made of alkylated derivatives of Sephadex (e.g. LH-20) and occasionally of rubber, Teflon and polyethylene as 'stationary phases'. It is not clear if the use of these hydrophobic materials with the range of solvents just described constitutes true partition chromatography or hydrophobic adsorption chromatography. However, the principles of partition chromatography appear to hold. These materials are resistant to bleeding with most solvents and after chromatography they can be stripped of any hydrophobic contaminants by washing the column with acetone, chloroform and pentane before reusing.

Probably this form of liquid chromatography is now most frequently used for qualitative, often preparative separations. HPLC is often the preferred form of liquid chromatography for analytical work and is increasingly the method of choice for preparative work also. The convenience and relatively low cost of TLC is sometimes preferred for qualitative analysis and in some special cases offers advantages for preparative chromatography.

High-performance (high-pressure) liquid chromatography (HPLC). HPLC is a more sophisticated logical development of LC (see *Figure 2.5* and e.g. refs 2 and 3). Very small support particles over a narrow range of defined size are used. This guarantees an even flow

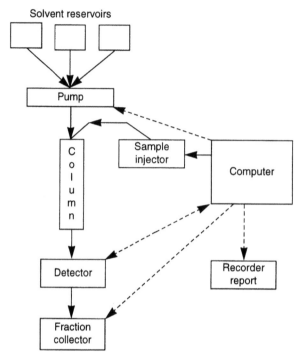

FIGURE 2.5: *Basic arrangement of equipment for HPLC.*

of solvent but requires a high pressure to move the solvent through at an acceptable and uniform rate. The particles carry a stationary phase that is chemically bonded to their surface (*Table 2.3*). The most commonly used stationary phase is a C_{18} hydrocarbon chain linked by condensation to silanol groups of a silica support. This octadecyl-silica (ODS or C_{18}) provides a reversed phase partition stationary phase. Other hydrophobic columns such as C_8- and phenyl-silica packings are also quite popular. Again whether or not this is truly partition chromatography or hydrophobic adsorption chromatography is uncertain. However, it provides columns of high resolving power (high values of N, low values of HETP) suitable for rapid quantitative chromatography. Chemically bonded phases suitable for normal phase partition chromatography include the cyanoalkyls (R-CN). In all of these and other chemically bonded silica based supports it is important to appreciate that different manufacturers sell stationary phases (usually as pre-packed columns) that differ in the proportions of bonded phase to silica, in the extent of end capping of unreacted silanol groups, and in the size (and range of size), shape and porosity of support particles. It is therefore essential to quote the name of the manufacturer and describe the product fully when reporting the use of HPLC.

TABLE 2.3: *Chemically bonded stationary phases commonly used in HPLC of lipids and related compounds in approximate order of increasing polarity*

Chemical type[a]	Structure[a]	Trade description
Octadecyl	$-(CH_2)_{17}CH_3$	ODS, C_{18}
Octyl	$-(CH_2)_7CH_3$	C_8
Phenyl	$-C_6H_6$	Phenyl
Cyano	$-CN$	CN
Amino	$-NH_2$	NH_2
Quaternary ammonium	$-N[(CH_2)_nH]_3$	$-SAX$ (anionic)

[a]Usually linked covalently to a silica support.

With regard to the mobile phases used, most are organic and, if a UV absorption detector is to be used, it is important to use a UV-transparent solvent such as hexane, methanol, isopropanol or acetonitrile. Many HPLC systems use detectors in the 200–210 nm range. Traces of solvent impurities can give rise to absorption of UV light, placing a premium on the use of highly purified solvents. Plastic-ware is best avoided for it contains UV-absorbing plasticizers and anti-oxidants which leach out. HPLC-grade solvents are usually free of these contaminants. It is also important to remember that silica-based packings are unstable outside the pH range 2–7.5.

Isocratic elution (mobile phase of constant composition) is sometimes satisfactory but it is usually more efficient bearing in mind the speed, capacity and sophistication of HPLC to employ a gradient of changing solvent composition designed to elute slow moving components earlier.

HPLC columns are intended for repeated use. This can present a problem if slightly dirty samples are injected on to the column, giving rise to an accumulation of contaminants, often resinous material, at the beginning of the column which tends to block it and reduce chromatographic efficiency. A small, replaceable, guard column of the same material preceding the main column provides a convenient, relatively cheap way of filtering out the 'dirt' without prejudicing the efficiency of the main separation column.

Standard HPLC columns are of 4.6 mm internal diameter (i.d.). Recently improved column packing technology has allowed the use of narrower columns. There are advantages to be gained in using narrower columns for analytical HPLC. These include greater sensitivity (up to 100-fold), greater speed and economy of using less (often 100-fold) solvent. Columns as narrow as 1 mm (often called microbore columns) are now available routinely pre-packed with the usual range of adsorbents/support materials. Fused silica capillary columns (320 μm i.d.) are also now added to the range of columns available. However they can be used only with equipment especially designed or modified to handle small volumes and narrow columns.

A critical factor in the success of HPLC has been the development of very sensitive flow-through detectors (*Figure 2.6*). The most commonly used are variable wavelength detectors. Their reasonably high sensitivity is rarely affected by changes in flow rate, solvent composition or temperature especially if a tapered flow cell is used. They offer the opportunity of selective assay of particular solutes that show absorption at specific wavelengths and also of general assay of most lipids which show absorption of light in the region 200–210 nm. This latter general assay is dependent upon using eluting solvents that are transparent to light of low wavelengths. This may restrict solvent use to those of less separative capacity. In some cases it is possible to overcome this problem by forming chemical derivatives of the solutes which absorb light at a higher wavelength where solvent mixes of high separative capacity are transparent; for example, esterification of alcohols with aromatic or colored acids (e.g. to form benzoates or diazobenzoates). Further examples are given in later chapters. However, this need to form a derivative nullifies one of the main advantages of HPLC over other chromatographic procedures such as GLC, namely the ability to chromatograph native lipid mixtures directly.

Fluorescence detection may be two or three orders of magnitude more sensitive than UV detection. However, in most cases it is necessary to synthesize fluorescent derivatives before this method can be used. On the other hand detection of changes in refractive index (RI) is often two to three orders of magnitude less sensitive than UV detection. Sensitivity depends on the difference in RI between solute and solvent as they pass through the RI detector. This is very sensitive to changes in temperature and will clearly vary if gradient solvent systems are used. It has been used widely in analysis of major lipid components such as triglycerides and fatty acids.

When used to follow carboxyl group absorption at 5.75 μm, infrared (IR) detectors have the same sensitivity as RI detectors especially if the solvent background can be controlled. They have been used to detect fatty esters, fatty acids and fatty aldehydes eluted by gradient solvent systems.

Flame ionization detectors (FID) are very sensitive and are used with high success in GLC (see the following section). They have had limited

(a)

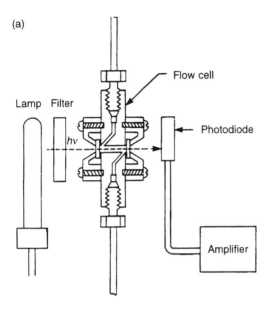

(b)

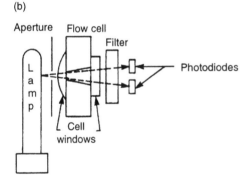

FIGURE 2.6:   *Diagrammatic representations of* (a) *a standard flow-through detector and* (b) *a tapered flow cell. The tapered cell ensures that changes in the refractive index of the solution passing through the cell, for example during gradient elution, do not lead to loss of light due to the light beam being deflected on to the walls of the cell. This design development results in a flatter base line on the elution profile. Redrawn from ref. 4 with permission from Academic Press.*

success in HPLC. Here the column eluent is first deposited on a moving wire and the solvent is evaporated. The remaining solute is detected on the moving wire by burning it to form carbon dioxide and water, the former then being reduced over a nickel catalyst to methane which is detected by FID as in GLC.

Radioactivity detectors are available for on-line assay of ^{14}C, ^{3}H and ^{32}P in lipid solutes at high sensitivity. Premixing of column eluent with scintillation fluid prior to assay of the light emitted in a flow-through photometer is the most sensitive method but may present problems if recovery of solutes is important. Alternatively, passing the radioactive eluent over a solid scintillant packed into a flow-through photometer may be preferred. In this case care has to be taken to avoid build-up of background due to continuous adsorption of traces of radioactive material on the scintillant.

Following the success of gas chromatography–mass spectrometry (GC–MS), systems have been developed that allow interfacing of HPLC with mass spectrometers. The revealing of structural detail of the solute as well as its sensitive quantification by mass spectrometry make this a very powerful technique. Further technical developments are extending the range of lipids amenable to structural analysis in this way. However, the cost remains high, limiting availability in laboratories for routine analysis.

Although several HPLC systems enable analysis of individual lipids directly on a total lipid extract many analyses benefit from a preliminary fractionation into lipid classes. As well as simplifying the chromatographic problem this may also remove interfering impurities and allow solvent systems to be used which are more amenable to using sensitive detection methods. Often this preliminary fractionation employs adsorption chromatography on silica gel, although anion exchange chromatography can be effective. A particularly convenient method is to use small Sep-Pak® columns of silica gel.

These mini-columns, currently available from the Waters Corporation (previously Millipore-Waters Chromatography) [5], are available as disposable, completely enclosed squat cylindrical cartridges with 100–500 mg high quality, high capacity silica of 1–3 ml volume, packed in tough, solvent resistant polyethylene tubes and retained by 35 µm polyethylene frits either end (*Figure 2.7*). They are suitable for direct attachment to syringes for supply of sample and eluting solvent under pressure. Alternatively Sep-Pak Vac cartridges come in a syringe-like arrangement with the upper barrel part empty and open with the same amounts of packing at the base of the barrel (*Figure 2.7*). They are designed to hold a reservoir of dilute sample or developing solvent in the empty barrel which is drawn through the silica by slight vacuum from the bottom to be collected in a vessel in a vacuum chamber. Typically in lipid analyses samples are loaded in hexane and eluted by hexane mixtures with ethyl acetate or acetone. Volumes used are small (10 ml or so) and the process takes just a few

Anatomy of a Sep-Pak Cartridge

A Precision Chromatographic Tool For Sample Enrichment And Purification

Here are ten unique features that set Sep-Pak Cartridges apart from the competition and make them the best solid-phase extraction cartridges available today.

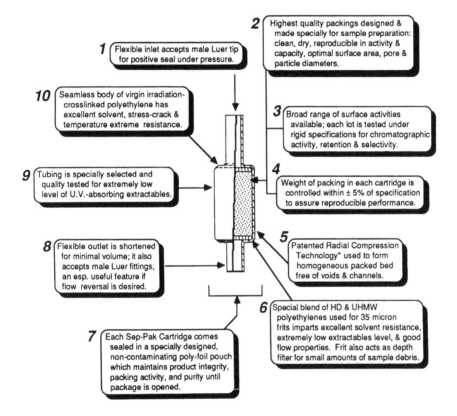

1 Flexible inlet accepts male Luer tip for positive seal under pressure.

2 Highest quality packings designed & made specially for sample preparation: clean, dry, reproducible in activity & capacity, optimal surface area, pore & particle diameters.

10 Seamless body of virgin irradiation-crosslinked polyethylene has excellent solvent, stress-crack & temperature extreme resistance.

3 Broad range of surface activities available; each lot is tested under rigid specifications for chromatographic activity, retention & selectivity.

9 Tubing is specially selected and quality tested for extremely low level of U.V.-absorbing extractables.

4 Weight of packing in each cartridge is controlled within ± 5% of specification to assure reproducible performance.

8 Flexible outlet is shortened for minimal volume; it also accepts male Luer fittings, an esp. useful feature if flow reversal is desired.

5 Patented Radial Compression Technology* used to form homogeneous packed bed free of voids & channels.

7 Each Sep-Pak Cartridge comes sealed in a specially designed, non-contaminating poly-foil pouch which maintains product integrity, packing activity, and purity until package is opened.

6 Special blend of HD & UHMW polyethylenes used for 35 micron frits imparts excellent solvent resistance, extremely low extractables level, & good flow properties. Frit also acts as depth filter for small amounts of sample debris.

* P.D. McDonald, R.V. Vivilecchia, D.R. Lorenz, Triaxially Compressed Beds, U.S. Patent 4,211,658 (1980); Australian Patent No. 509,338; other patents pending.

FIGURE 2.7: *A manufacturer's description of the advantages of using Sep-Pak cartridges. Reproduced from ref. 5 with permission from the Waters Corporation.*

minutes. The capacity of these cartridges ranges from a few milligrams of lightly retained to 100 mg of strongly retained lipids. They can also be used to retain lipids present at dilute solution in large volumes of nonpolar solvents which are then eluted from the cartridges in small volumes of more polar solvents, thus concentrating the solution.

Sep-Pak cartridges containing reversed phase packings and ion exchange packings have also found use in lipid sample preparation protocols. Their use is modeled on the principles of reversed phase and ion exchange LC discussed elsewhere in this section, adapted to the advantages of high capacity and high speed outlined above for the silica packing.

Gas–liquid chromatography (GLC). GLC involves the separation of volatilized components of a mixture as they are passed in a stream of inert carrier gas (mobile phase) over a layer of nonvolatile liquid (stationary phase) coated on a solid support, usually of small particles, packed in a column (see *Figure 2.8* and e.g. refs 6 and 7). Characteristics of the compounds which influence particularly their separation are their relative volatilities and solubilities in the stationary phase. Samples are usually injected as solutes in a volatile organic solvent (1–5 µl) directly on to the column preset at a temperature high enough to volatilize solvent and samples. The flow of carrier gas removes the more volatile solvent as vapor rapidly down the column. The less volatile solutes are retained by the stationary phase to different extents and so move along the column at different rates to be eluted from the column at different times.

Clearly GLC can only be used to separate and assay lipids that are volatile or can readily, preferably quantitatively, be converted to a

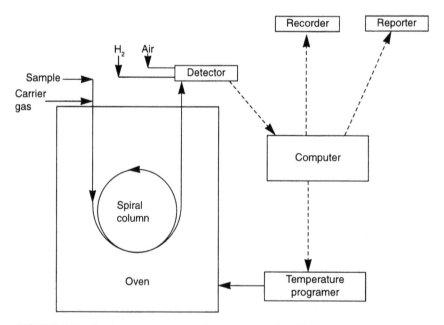

FIGURE 2.8: *Basic arrangement of equipment for GLC.*

derivative which is volatile at temperatures below 350°C. Some of the largest lipid molecules analyzed successfully by GLC are triacylglycerols which are volatile without derivatization at this temperature. Most lipid components which are more polar than triacylglycerols require conversion to a volatile derivative before GLC can be used. For example, unsubstituted hydroxyl groups can be substituted with trimethyl silyl groups or acetylated and carboxylic acid groups can be methylated. Other examples are detailed in later chapters.

Stationary phases generally fall into one of two groups: polar or nonpolar. Some are listed in *Table 2.4*. Appropriate selection of a stationary phase can ensure very efficient resolution of the components of mixtures. The advantages of one phase over another

TABLE 2.4: Stationary phases commonly used in GLC of lipids and related compounds (listed in approximate order of low to high polarity)

Chemical type	Commercial names	Analytical uses
Saturated hydrocarbon	Apiezon-L or -M	
Methylsilicone	CP-Sil5, DB-1, DC-200, HP-1, OV-1, OV-101, SE-30, SF-96, SP-2100, Rtx-1	Hydrocarbons, fatty acids (methyl esters), long chain alcohols. Sterols, steroid hormones. Chlorinated hydro-carbons. Triglycerides, TMS sugars
Methylphenylsilicone (5–10% phenyl)	CP-Sil 8, DB-5, Dexsil 300, DC-560, Fluorolube, HP-5, OV-3, OV-5, OV-73, PTE-5, SE-52, SE-54, SP 2250, Xti-5	
Cyanopropylphenyl silicone (6–14%)	CP-Sil 19, DB-1301, DB-1701, HP1301, OV-1701 Rtx 1301, Rtx 1701, Silar 10CP, SP2330	
Methylphenylsilicone (35–50% phenyl)	DB-17, DB-35, DC-710, HP-50, OV-11, OV-17, OV-22, OV-25, Rtx-50, SP-2250, SPB-35	
Trifluoropropyl	DB-210, OV-202, OV-210, OV-215, Rtx-260, SP-2401, QF-1,Triton X-100, UCON HB 280X	Unsaturated fatty acids (methyl esters), aldehydes, alcohols and hydrocarbons. Short chain fatty acids (free). Monosaccharide derivatives
Polyethylene glycols (and derivatives)	AT-1000, Carbowax-20M, CP-wax 51, DB-wax, DEGS, ECNSS, EGSSX, FFAP, HP-wax, HP-Innowax, OV-351, PEG, Supelcowax-10	
Cyanopropyl silicone (50%)	CP-Sil 58, DB-23, DEGS, SP2310, SP2330, OV-275	

depend on the particular analysis to be undertaken. Examples of successful applications are discussed in more detail in several of the following chapters.

The time of elution of solutes from a particular GLC column is very sensitive to change in temperature. For this reason the temperature of most GLC columns is controlled by placing them in an oven fitted with a thermostat accurate to 0.05–0.1°C. Increase in the speed of elution of solutes can often be achieved by carefully controlled (usually programed) increases in oven temperature. The use of a temperature gradient may be an important aspect of completing a successful GLC analysis within a minimum time period.

Several reviews of the theory of GLC have been published (see e.g. refs 6 and 7). Most features of GLC theory can be described adequately by extending the general discussion in Section 2.2.1. Since most GLC systems involve passage of quite large volumes of carrier gas and at uniform flow rate, it has become common practice to use elution time and not elution volume (*Figure 2.4*) when analyzing patterns of elution. A standard arrangement of GLC equipment is illustrated in *Figure 2.8*.

Columns are usually 1.5–2.5 m long coils of 2–4 mm diameter glass or stainless steel tubing. Gas flow (often argon or oxygen-free nitrogen) is kept constant and is usually of the order of 50 ml min^{-1}. Most standard columns packed with particles coated with stationary phase at 10–15% (w/w) will maintain an efficiency of at least 3000–5000 theoretical plates for a year or more.

Capillary columns have become popular for analytical work. They normally consist of long coils of narrow bore tubing (e.g. 100 m × 0.25 mm i.d.), the inner wall of which is coated with the stationary phase. Sample application is usually in a small volume (10^{-2} μl). Efficiencies of 20 000–100 000 theoretical plates are often achieved, giving excellent separations of even closely similar compounds (e.g. isomers). They are the columns of choice for GC–MS. Shorter capillary columns giving less resolving power are sometimes used if very rapid analysis (over a few seconds) of mixtures of easily separable components is required.

Routinely, detectors for quantification in GLC are of three types: the thermal conductivity detector (katharometer), the FID and the electron capture detector (ECD) (*Figure 2.9*). In addition the mass spectrometer is being used increasingly for quantification and structural analysis of GLC eluents.

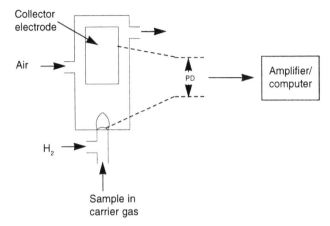

Air

H₂

Collector electrode

PD

Amplifier/ computer

Sample in carrier gas

FIGURE 2.9: *Basic aspects of a flame ionization detector (FID). PD: potential difference between sample/carrier gas inlet to flame and charge collector electrode.*

The FID is the most popular detector. This consists of two platinum electrodes across a flame of hydrogen burning in air. The carrier gas leaving the column is introduced into the flame and molecules ionized in the flame generate a current across the electrodes. This arrangement can detect most organic compounds with a detection limit of 10^{-9} mol and a linear response. The alkali flame ionization detector (AFID) is a version of FID in which the presence of alkali atoms (K, Ru or Cs) in the flame enhances sensitivity to nitrogen- and phosphorus-containing compounds about 10-fold.

The simplest detector is the thermal conductivity detector which involves measurement of difference in the electrical resistance of a piece of platinum wire placed in the flow of gas leaving the column compared with another piece in the flow of carrier gas only. Both pieces of wire are heated electrically. The resistance of the wires varies with changes in their temperature. This in turn depends upon heat loss from the wires which varies with the composition of the gases passing over them. Sensitivity is about 10-fold less than that of the FID and the response is less linear.

Of the three types of detectors, ECDs are the most sensitive, detecting as little as 10^{-12} mol, but they are specific for electrophilic compounds. In these detectors, β-particles from a radio-isotope (e.g. ^{63}Ni or ^{3}H) ionize the carrier gas as it passes through the detector and a current is developed. An electrophile in the gas from the column will capture some of the electrons in the ionizing gas and so reduce the current. Detector response is usually quite linear.

Paper chromatography and thin-layer chromatography (TLC).
Partition chromatography employing a stationary phase of water
bound to cellulose in the form of sheets of paper or thin layers of
particles adhering to a thin plate of glass or other flat inert material
has been popular for many years, (see e.g. ref. 8). The mobile phase in
these systems is less polar than water (*Table 2.2*) and flows up or
down the sheet or layer by capillarity, usually for a predetermined
distance (*Figure 2.1*). TLC has been especially popular with lipid
analysts.

For most analytical TLC, layers of adsorbent/stationary phase support
of 0.2 mm thickness is satisfactory. For preparative work much
thicker layers may be used. Successful chromatography can usually
be achieved across/along 15 cm or so of stationary phase leading to 20
× 20 cm square plates being the most common. Most chromatography
tanks are designed to accommodate this size. Variations on this size
are available commercially. In particular microplates (1 × 3 in) are
convenient for rapid monitoring of column LC systems for the elution
of one or more solutes.

Both precoated and noncoated glass plates are available
commercially. In the laboratory noncoated plates may be covered by a
uniform layer of adsorbent by using simple equipment designed to
spread evenly an aqueous slurry of the adsorbent over several plates
and then drying them in an oven at 100–120°C. The presence of a
binding agent such as calcium sulfate in the slurry ensures that the
dried adsorbent sticks well to the plate. Stable layers of adsorbent are
also available commercially on a plastic or aluminum foil backing.
These have the advantage of flexibility and ease of cutting to suit
nonstandard chromatographic arrangements and recovery of solute
on small pieces of the adsorbent layer.

Samples are usually applied to TLC plates as solutions, preferably in
a solvent of moderate volatility and of low polarity and by
micropipette or syringe. Application is of a few microliters to form a
small spot at a point about 2.5 cm from one end of the TLC plate. The
evaporation of the solvent may be assisted by an air blower.
Alternatively a thin line of sample at right angles to the intended
direction of chromatographic development may be applied, especially
for preparative purposes. Care must be taken not to damage the layer
of adsorbent or in any way risk interfering with the subsequent flow
of chromatographic solvent. To maintain a small application spot or
narrow line of a dilute solution, multiple applications and drying may
be necessary. Polar solvents, especially if used for multiple
applications, may result in deposition of the solute at the
circumference of the spot or primarily as two narrow lines;

arrangements which may lead to distorted chromatograms. Less polar solvents should ensure a more even distribution of sample across the spot/line. Some companies sell TLC plates with a 3 cm strip of a weakly preadsorbent layer at one end to which the sample application is made. This causes each spot to be compressed into a narrow horizontal line (or each double line into one narrow line) at the junction of the two adsorbents during the chromatography, so obviating this problem.

Chromatography is started by placing the plate in a chromatography tank which contains the developing solvent at a depth of about 1.5 cm and which has been saturated with solvent vapor during 1 h or so with the lid tightly on (possibly assisted by solvent-saturated chromatography paper around the inner wall of the tank). The plate is positioned almost vertically with the sample(s) at the bottom. The developing solvent travels up the layer of adsorbent by capillarity, dissolving the sample(s) as it does so. To avoid anomalous movement of the sample(s) due to uneven evaporation and movement of the solvent, it is essential to keep the system draught-free and at a constant temperature and to ensure a tight seal between the lid and the tank. The solvent may take between 20 min and 2 h to reach a preselected line near the top of the plate depending on the viscosity, density and surface tension of the solvent, the temperature and the vertical distance from the point of application (the origin) to the preselected line (the solvent front).

In most areas of lipid analyses, reversed phase has been the most useful form of partition thin-layer chromatography. At its simplest this has involved preparing or purchasing TLC plates of kieselguhr or silica gel and immersing them in a dilute solution of the stationary phase (e.g. liquid paraffin) in a volatile solvent. The plates are then allowed to dry free of solvent resulting in an even distribution of stationary phase. Mobile phase is saturated with stationary phase at the temperature at which the chromatography is to be carried out. TLC plates carrying hydrophobic groups linked chemically to a layer of silica (see *Table 2.3*) are sometimes used. They are more expensive than 'home-made' reversed phase plates but are much easier and less messy to use. In these cases mobile phases do not need to be saturated with stationary phase.

It is generally regarded that most TLC systems show separation efficiencies of between 1500 and 5000 theoretical plates, slightly less than most HPLC and GLC systems but greater than normal LC. Paper chromatography and TLC have the advantage that poorly resolved samples can be chromatographed in a different solvent system at right angles to the first run if square or almost square paper (plates) is used (see *Figure 2.10*). This two-dimensional approach enables a solvent system appropriate for fast running components to be used in one dimension and for slow running components in the second dimension.

Other advantages of TLC are its relative cheapness, simplicity and speed (especially if multiple samples are run in parallel on the same plate). The disadvantages are the small capacity and the difficulty of accurate quantification.

Components separated by TLC or paper chromatography are deposited at different points along the planar chromatogram. Characterization of most components will depend upon relative chromatographic mobility, preferably determined in several TLC and/or paper chromatography systems and where possible by co-chromatography in these systems with authentic material putatively identical to the unknown component. Identification may also be aided by sensitivity to particular stains and/or absorption of UV or visible light.

Chromatographic mobility is usually described in terms of the distance traveled by the component from the point of application relative to the distance traveled by the solvent front from the same point (see *Figure 2.11*); this is known as the R_f value. Tables of R_f values of lipids in different defined systems are published in the literature. If the system requires the solvent (mobile phase) to travel

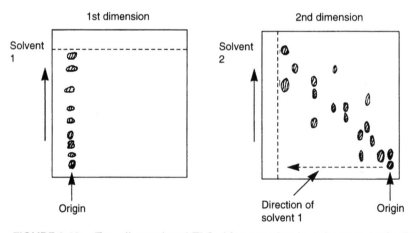

FIGURE 2.10: *Two-dimensional TLC. After running in solvent 1 vertically the TLC plate is allowed to dry before rotating it through 90° anti-clockwise and running in solvent 2 vertically.*

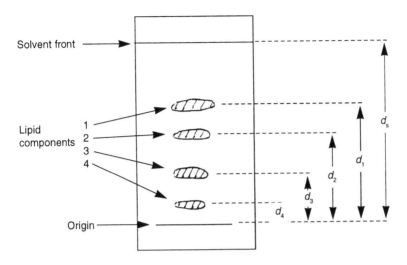

FIGURE 2.11: R_f values in TLC. The distance traveled up the plate of the separated components is related to the distance traveled by the solvent front measured from the origin (point of application of the lipid mixture). The R_f values of components 1–4 will be d_1/d_s, d_2/d_s, d_3/d_s and d_4/d_s, respectively.

beyond the end of the sheet or plate (by allowing evaporation at that end) mobility is related to the distance traveled by a standard compound alongside or the R_s value, where s is a named standard.

Detection of lipids and derivatives that contain no chromophoric or characteristically reactive group, as is the case with many lipids, is usually by charring through heating after spraying with sulfuric acid or dichromate reagent, by staining with iodine or by spraying with a fluorescent dye. Quantification usually requires careful calibration of these procedures and may be difficult. This may involve densitometry of the stained chromatogram or removal of the stained region and assay of the intensity of the stain in a photometer.

2.2.3 Adsorption chromatography

Adsorption chromatography depends on differences in the balance between the solubility of solutes in a mobile solvent and their adsorption to the surface of small solid particles over which the solution is passed. Usually the adsorbent solid is polar and may sometimes be acidic or basic. The adsorption of solutes to polar solids probably involves hydrogen bonds. Some adsorbents are nonpolar, in which case hydrophobic bonds between solutes and adsorbent are important. A small number of polar solids are used routinely, most commonly silica and alumina. However, a large number of solvent mixtures have led to many adsorption chromatographic systems. In these systems solvents are selected that maximize the differences in mobility between components of the mixture of solutes that are of particular interest. Strategies for achieving this differ in LC, HPLC and TLC.

In classical column adsorption chromatography on alumina or silica stepwise increases in polarity of the eluting solvent are usually arranged by judicious use of appropriate mixtures of solvents using the information in *Tables 2.1* and *2.2*. In HPLC a similar aim is achieved by using the pumping arrangement (see *Figure 2.5*) to deliver a continuously changing, linear gradient, rather than a stepwise gradient. In TLC one is usually restricted to isocratic chromatography. Here the single solvent or mixture of solvents is usually chosen to maximize separation of the components of interest, accepting that the separation of other components may be minimal. Two dimensional TLC may provide a means of separating most of the components on one TLC plate.

The adsorptive power of alumina and silica is reduced by increased hydration, offering a further possibility of influencing chromatographic separations. It has been common to use silicic acid (silica) for separations of glycerol-based lipids and alumina for isoprenoids.

The conventions for describing and assessing partition chromatographic separations are also used in adsorption chromatography, especially in HPLC and TLC. The arrangements of equipment and methods of detection and assay used are almost identical in corresponding types of partition and adsorption chromatography. The reader is referred to Section 2.2.2 for a detailed consideration of these aspects.

2.2.4 Ion exchange chromatography

Classical chromatography of lipids on synthetic polymers that carry ionic groups depends primarily upon binding of these groups to groups on the lipids of opposite ionic charge. Thus a polymer with fixed cations will bind anionic lipids from a solution providing that the pH of the solution ensures ionization of the anionic groups and that the concentration of nonlipid anions in solution is not sufficient to compete for all of the fixed cations. In fact the interaction of the lipid with the polymer may be more complex than this, involving, for

TABLE 2.5: Ion exchange chromatography materials commonly used in lipid analysis

Ionizing group[a]	Trade description	Analytical uses
$-(CH_2)_2NH(C_2H_5)_2$ diethylaminoethyl	DEAE (anionic exchanger)	Anionic lipids (phospho-, sulfolipids, sialoglycolipids, fatty
$-(CH_2)_2N(C_2H_5)_3$ triethylaminoethyl	TEAE– (anionic exchanger)	acids)
$-CH_2COO^-$ carboxymethyl	CM– (cationic exchanger)	Phospholipid mixtures

[a]Usually linked covalently to a cellulose support.

example, adsorption of polar nonionic groups by hydrogen bonds, especially if the lipid contains carbohydrate or inositol.

The most common form of ion exchange chromatography for polar lipids has employed columns of the anion exchanger diethyl-aminoethyl (DEAE) cellulose usually in the acetate form (*Table 2.5*). It is most effective over a pH range of 3–6. Often elution is achieved by stepwise increases in concentration of the buffer ammonium acetate in water–alcohol mixtures. The cation exchanger carboxy-methyl cellulose in the Na+ form has been used occasionally over the pH range 7–10 for phospholipid separations.

Bulk anion exchange chromatography can be effective for preliminary separation of anionic lipids from uncharged and zwitterionic lipids. Several systems have been described. These have employed DEAE Sephadex A-25, DEAE Sepharose (Fast Flow or CL-6B) or Q-Sepharose as well as DEAE cellulose. The eluting solvent is usually a gradient of increasing concentration of ammonium acetate dissolved in a mixture of chloroform–methanol (1:2, v/v). The separation may be best monitored by TLC.

In HPLC versions of ion exchange it is important that particles bearing the fixed charge are very small, of uniform size and charge distribution and are physically stable to the high pressures developed in the chromatography column. DEAE cellulose and related materials cannot meet these criteria. Currently high-performance anionic exchange chromatography (HPAEC) is achieved through chemically linking amine groups to silica via a paraffinic linker such as occurs in μ-Bondapak-NH_2. HPAEC on columns of this material offers good resolution and sensitivity by on-line UV detection at 210–215 nm. For example, a gradient of aqueous phosphate buffer in acetonitrile has been employed successfully for ganglioside separations at sub-nanomolar levels using this column packing [9].

Ion pair chromatography can be viewed as a form of high-performance reversed phase partition chromatography or of ion exchange chromatography. It employs a hydrophobic HPLC column such as C_{18} and the mobile phase contains a relatively hydrophobic counter-ion to the ionic materials to be separated. For example, tetrabutylammonium (Bu_4N^+) will form a neutral ion pair (salt) with anions which can then be seen as undergoing reversed phase partition chromatography as a complex. Alternatively the tetrabutylammonium can be seen as binding hydrophobically to the stationary phase leaving a positive charge on its surface and producing an anion exchanger. Most chromatographers prefer the former rationalization of the phenomenon. Whichever is correct, the method has been used successfully for phospholipid separations.

The arrangement of equipment used for ion exchange chromatography of lipids is usually that used in partition chromatography and is covered in Section 2.2.2.

2.2.5 Complexing chromatography

There are a small number of examples of the use of immobilized compounds or of compounds/ions in solution that form a complex with specific groups in a lipid and so modify the chromatographic mobility of the compounds concerned without influencing that of the other lipids present. These examples include silver nitrate which interacts reversibly with double bonds in aliphatic chains. When incorporated into adsorbents it enables separations on the basis of number, position and *cis* and *trans* isomerism of double bonds in unsaturated fatty acids and their derivatives by adsorption LC or TLC. A second example is the use of borate which complexes with compounds containing hydroxyl groups on adjacent carbon atoms, assisting the separation of glycolipids.

2.3 Spectrometric methods

Ultraviolet (UV) and visible spectrometry can be used both for the identification of lipids and for their quantification. A purified sample is desirable for the latter. Other methods for the characterization of lipids are infrared (IR) and nuclear magnetic resonance (NMR) spectroscopy and mass spectrometry (MS).

2.3.1 UV–visible spectrometry

The books by Dyer [10] and Crooks [11] should be consulted for the underlying principles and more detailed information on this technique. Solutions of lipids may absorb visible or UV light at characteristic wavelengths. The absorbance, A, or optical density is defined as:

$$A = \log \frac{I_0}{I}$$

where I_0 is the intensity of incident light and I is the intensity of transmitted light. Absorbance is usually recorded in the range 0–2.0. The Lambert–Beer law defines the molar extinction coefficient, ε, as follows:

$$\varepsilon = \frac{A}{c\, l}$$

where c is the molar concentration and l is the path length of light through the sample (usually 1 cm). The lipid solution is placed in a glass cell for studies in the visible region (400–800 nm), or in a quartz cell for the UV region (180–400 nm). Suitable solvents are hexane, cyclohexane or 95% (v/v) ethanol and the lipid concentration is usually about 10^{-4} M.

Certain functional groups such as carbonyl or double bonds absorb UV light strongly at certain wavelengths, so the best aid to lipid identification is a scan of the absorption over a range of wavelengths. *Table 2.6* lists the wavelengths (λ_{max}) and ε_{max} of some chromophoric groups.

Modern spectrometers have a built-in computer and display screen as well as a printer. The Philips PU8700 series of instruments, for

TABLE 2.6:   *UV absorption characteristics of some chromophoric groups[a]*

Chromophore	Example	λ_{max}(nm)	ε_{max}	Solvent
–C=C–	Octene	177	12 600	Heptane
–C=C–C=C–	Butadiene	217	20 900	Hexane
C_6H_6	Benzene	184	47 000	Cyclohexane
		202	7000	
		255	230	
CHO	Acetaldehyde	290	17	Hexane
COOH	Acetic to palmitic acids	208–210	32–50	Ethanol
–CH=CH–COOH	*cis*-Crotonic acid	206	13 500	Ethanol

[a] Selected from ref. 1 with permission from the author and Elsevier Science.

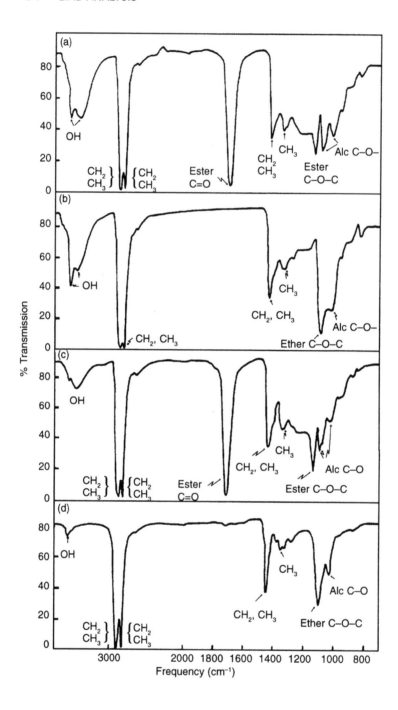

FIGURE 2.12: *IR spectra of:* **(a)** *1-monostearin;* **(b)** *batyl alcohol;* **(c)** *1,2-distearoyl glycerol;* **(d)** *1,2-di-O-octadecyl glycerol. Alc, alcohol. Adapted from ref. 1 with permission from the author and Elsevier Science.*

instance, are operated from the screen using a 'mouse'. Data can be stored and recalled for further manipulation.

2.3.2 IR spectroscopy

A fuller account can be found in ref. 1. Molecular stretching or bending vibrations give rise to many absorption bands in the IR region, which covers the wavelengths 2.5–15 µm. Bands are usually defined by wavenumber (frequency), i.e. the number of waves per centimeter.

Since water has strong absorption bands, the lipid sample must be carefully dried in a vacuum desiccator. Glass and quartz also absorb strongly, so alkali halides are used. Oils can be examined as thin films between sodium chloride plates. Other lipids can be dissolved in carbon tetrachloride or chloroform at 10–20 mg ml⁻¹ and cells with sodium chloride windows used. Solid lipids which are not sufficiently soluble in these solvents can be made up in potassium bromide pellets. About 2 mg of sample is ground finely with 200 mg KBr, dried and pressed into a transparent disc at high pressure. The disc is placed in the holder of a double-beam IR recording spectrometer and the spectrum recorded in the region 4000–600 cm⁻¹.

A table of absorption bands due to various functional groups found in lipids is given in ref. 1. Some typical IR spectra are shown in *Figure 2.12*. The absorption bands associated with the phosphate groups of phospholipids vary with the ionic form of the phospholipids.

2.3.3 NMR spectroscopy

NMR spectroscopy is a powerful technique and taken together with IR spectroscopy can provide a great deal of information on lipid structures. Dyer [10] gives a fuller account.

Nuclei of protons and certain isotopes such as ^{13}C and ^{31}P behave as minute magnets and will resonate when irradiated with radio waves in a strong magnetic field. Positions of absorption bands (chemical shifts) are given relative to tetramethylsilane as a standard. The parameter δ (delta) is used, in units of parts per million (p.p.m.). It is defined as

$$\delta = \frac{\Delta\upsilon \times 10^6}{\text{oscillator frequency (in Hz)}}$$

where $\Delta\upsilon$ is the difference in absorption frequencies of the sample and standard (in Hz).

Sample preparation is similar to that for IR spectroscopy and fairly concentrated solutions are used. For proton resonance (PMR) spectroscopy, carbon tetrachloride is a suitable solvent and for ^{13}C NMR deutero-chloroform may be used. Examples of NMR spectra and tables of δ values for various structural groups found in lipids are given in ref. 1. In PMR spectra, the total integrated area under the peaks is proportional to the total number of hydrogens in the molecule studied. Thus the proportion of hydrogens in any particular functional group is given by the area under the relevant peak or peaks divided by the total area under all peaks. The PMR spectrum of methyl phytanate is shown in *Figure 2.13*.

2.3.4 Mass spectrometry

The book by Pavia *et al.* [12] gives an introduction to MS and to other spectroscopic methods dealt with in the previous sections. Kates [1] deals with applications of MS to lipids.

In the mass spectrometer molecules are bombarded by a stream of high-energy electrons and the resulting ions are separated in a magnetic or electric field according to their mass-to-charge ratio. A detection device then counts the number of ions with a particular mass-to-charge ratio. The resulting spectrum has a pattern of peaks representing molecular fragments and the 'parent peak' due to the unfragmented ionized molecule. Lipids such as phospholipids which

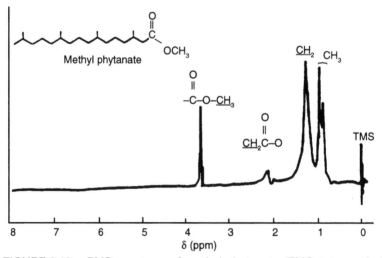

FIGURE 2.13: *PMR spectrum of methyl phytanate. TMS, tetramethylsilyl. Redrawn from ref. 1 with permission from the author and Elsevier Science.*

have low volatility and are unstable to heat, give no parent ion peak by conventional MS. They can be analyzed however, by techniques such as chemical ionization MS, where ionization and volatilization techniques are less drastic.

In the chemical ionization method, methane is introduced into the ionization chamber, along with the sample to be analyzed. Most of the electrons then ionize methane rather than sample molecules, with ions such as CH_5^+ being formed. Such ions react with sample molecules, for example:

$$CH_5^+ + RH = RH_2^+ + CH_4.$$

Such protonated (RH_2^+) molecules are accelerated in the usual way, giving peaks one mass unit higher than those expected in the normal electron impact MS. Other related techniques such as fast atom bombardment, secondary ion and field desorption MS are outlined in ref. 1.

Combined gas-liquid chromatography–mass spectrometry (GLC–MS) apparatus is available (see previous section on Gas–liquid chromatography). This is convenient since individual components of a mixture are separated by GC and automatically introduced into the mass spectrometer as they emerge.

A good deal is known about the ways in which different molecules fragment in the mass spectrometer. For instance, alcohols break near the hydroxy group with the elimination of water, while carbon-branched or keto compounds break on either side of the branch or keto group. Details are given in ref. 12. The mass spectrum of methyl stearate is shown in *Figure 2.14*.

Since many lipids occur in living tissues as families of molecules differing only in their fatty acid composition, MS is useful in analyzing the individual molecular species.

2.4 Radioisotopic methods

2.4.1 Choice of radioisotope

The availability of sensitive and accurate radioisotope assay systems has made it possible to detect and measure very small amounts of radioactive lipids (see e.g. refs 13 and 14). These may have resulted from *in vivo* biosynthetic studies, from *in vitro* enzymic assays or

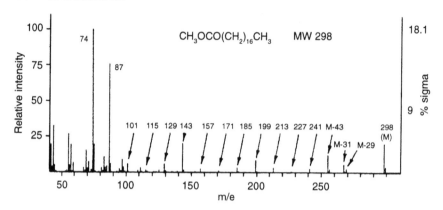

FIGURE 2.14: *Mass spectrum of methyl stearate. m/e, mass–charge ratio. Redrawn from ref. 1 with permission from the author and Elsevier Science.*

radioimmunoassays (RIAs) or from chemical formation of a radioactive derivative as part of a micro-assay of a nonradioactive lipid.

The most common radioisotopes used in lipid work are listed in *Table 2.7*. As they decay these unstable isotopes give off β-radiation in which the β-particles are emitted with a range of energies up to a maximum value ($E_{\beta max}$). The energy is sufficient to darken a photographic plate or emulsion giving rise to the technique of autoradiography for radioisotope detection. The low energy emitters, especially ^{3}H, are best measured by liquid scintillation counting (LSC). By this method, ^{3}H can be measured readily with an efficiency of 60% and ^{14}C and ^{35}S with an efficiency of approximately 95%. [32]P can also be assayed efficiently this way, although the higher energy of its β-particles allows other assay systems to be used, for example proportional counting and Geiger counting. [32]P may be assayed in the presence of ^{3}H, ^{14}C or ^{35}S by Cerenkov counting, a simple form of LSC. These radioisotopes are often incorporated into lipids, either biologically or chemically, in place of the corresponding natural nonradioactive stable isotope with very little change in the properties

TABLE 2.7: *Radioisotopes commonly used in lipid analyses*

Name	Symbol	$E_{\beta max}$ (meV)	Half-life
Tritium	[3]H	0.018	12.3 years
Carbon-14	[14]C	0.156	5736 years
Phosphorus-32	[32]P	1.71	14.3 days
Sulfur-35	[35]S	0.167	87.4 days

of the lipid other than its becoming radioactive. If very radioactive, the lipid may be less stable chemically due to the degradative effect of the energy released.

The γ-emitting radioisotopes ^{125}I and ^{131}I are also used in lipid assays involving estimates of degrees of unsaturation and in RIA. These have decay energies of 0.036 and 0.364 MeV and half lives of 59.6 and 8.05 days, respectively.

2.4.2 Autoradiography

Autoradiography is the simplest method for detecting radioactivity distributed across a flat surface. It has been applied to histological sections of tissue, to paper and thin layer chromatograms and in dot blot analysis. The method depends on the radiation darkening photographic (X-ray) film or emulsion placed in close contact with the sample. The film or emulsion is then developed in the usual way. The chemistry of the process is complex but basically the β-radiation interacts with the photographic emulsion to produce electrons which reduce silver halides to small particles of metallic silver.

At its simplest, dry thin layer or paper chromatograms are clamped into close contact with X-ray film, often sandwiched between glass plates (*Figure 2.15*). The operation is conducted in the dark and the chromatogram plus film is left in a light-proof container for an appropriate time. Standard X-ray film will usually require approximately 48, 16 and 8 h respectively to detect ^{3}H, ^{14}C (or ^{35}S) and ^{32}P when present on a surface at 100 Bq cm^{-2}. Absorption of radioactive material from the surface into a layer of paper, silica or alumina will increase these times. The use of high-performance autoradiography film (e.g. Hyperfilm ^{3}H, Amersham) may allow these times to be reduced substantially.

The detection of ^{3}H may be speeded up by using fluorography in which the chromatogram is impregnated with a scintillant sensitive to β-radiation and the film is darkened by the resultant photons of light. The most common scintillant is diphenyloxazole (PPO), a chloroform solution of which is sprayed on to the chromatogram. After drying, the chromatogram and film are put together as before and stored at −70°C. At this temperature the light output of PPO is increased and the X-ray film performs better. In this way the detection time for 100 Bq cm^{-2} may be reduced to the order of 1 h. The use of 'pre-flashed' X-ray film improves the linearity of response to quantity of radioactivity present, which is essential if the technique is to be used quantitatively.

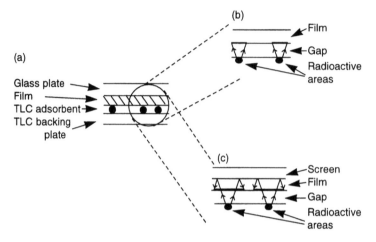

FIGURE 2.15: *Diagrammatic representation of autoradiography. (**a**) The TLC plate carrying radioactive components (dark areas) in the layer of adsorbent is placed in close contact with photographic film. A glass plate over this helps to maintain a good even contact between film and TLC plate. (**b**) Standard autoradiography in which β-radiation irradiates from the radioactive regions of the TLC plate to darken the adjacent region of the film. The smaller the gap the less broadening of bands and loss of resolution in the darkened film. (**c**) Autoradiography of high energy emitters (^{32}P and ^{125}I) showing the use of an intensifying screen to reflect the radiation back into the film as light. Because of the greater distance traveled by the radiation before darkening the film there is risk of greater broadening of bands and lack of resolution.*

In the case of ^{32}P and ^{125}I the high energy particles will pass through photographic film and are not detected. This can be countered by placing an intensifying screen of dense inorganic scintillator (e.g. calcium tungstate) behind the film. This converts radiation energy to light energy and results in superimposition of a photographic image on the radiographic image. This gives a 10- to 15-fold increase in sensitivity but inevitably the increased distance from the radioisotope to the film via the intensifying screen gives a loss of resolution.

Autoradiography and fluorography are relatively simple and cheap detection methods and with care may be used quantitatively (see also Section 2.5.3). Recently more sophisticated and expensive equipment has become available in which film is replaced by a phosphor screen with a reputed 100-fold gain in sensitivity for ^{14}C, ^{35}S, ^{32}P and ^{125}I. The screen can be used repeatedly and requires no chemical processing or cooling. In the Phosphor Imager (Molecular Dynamics) quantification and storage of the image is easily carried out.

2.4.3 Scintillation counting

Several substances (scintillants) emit a pulse of light when activated by a pulse of β-radiation. The amount (number of pulses) of light emitted is proportional to the amount (number of pulses) of radiation falling on the substance and can be measured accurately using a photomultiplier tube. The energy of the electrical pulses so generated is related to the energy (wavelength) of light produced, which in turn varies with the energy of β-radiation. With appropriate electronic discriminators of electrical energy, it is possible to set an instrument to measure several radioisotopes at the same time.

Most radioassay of this type involves mixing a solution of scintillant(s) with a solution or suspension of radioactive material and is described as liquid scintillation counting (LSC). A cocktail that has been popular for many years is of 2,5-diphenyloxazole (PPO) and 1,4-bis-2-(5-phenyloxazolyl)-benzene (POPOP) in toluene. In this cocktail, β-radiation causes a molecule of toluene to emit light of short wavelength which activates a molecule of PPO, the primary scintillant, to emit light of longer wavelength, which in turn activates a molecule of POPOP to emit light of an even longer wavelength and which corresponds to the maximum sensitivity of the photomultiplier (PM) tube (*Figure 2.16*). POPOP is insensitive to the radiation from toluene, hence the need for PPO. However, several alternative single scintillants are now on the market [e.g. 2-(4'-*t*-butylphenyl)-5-(4"-

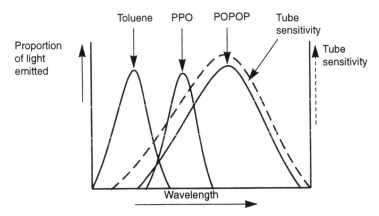

FIGURE 2.16: *The use of scintillants PPO and POPOP to optimize the matching of the light emission of toluene, having been stimulated by β-radiation, to the wavelength sensitivity curve of photomultiplier tubes of scintillation counters. The low wavelength emission of toluene stimulates PPO to emit light of longer wavelength which, in turn, stimulates POPOP to emit light of longer wavelength that matches well the tube sensitivity.*

biphenylyl)-1,3,4-oxadiazole, also known as butyl-PBD] which are sensitive to the emission of stimulated toluene and emit light at a wavelength appropriate for the PM tube. Most lipids dissolve readily in toluene and so LSC is a very convenient method for these. Some polar lipids are less soluble in toluene but are soluble in polar solvents that are immiscible with toluene. In this case, the detergent Triton X-100 may also be used to stabilize fine emulsions with toluene. Alternatively, a dioxane-based scintillation fluid may be suitable.

The advantages of LSC over other methods of radioassay include:

(i)   rapidity of response and response decay which allows many events per second to be counted, i.e. accuracy over a wide range of quantity of radioisotope;
(ii)   high efficiency of counting;
(iii)   ability to count several radioisotopes concurrently;
(iv)   ability to assay a wide range of sample type given appropriate selection of solvent and scintillant.

A minor disadvantage is the possibility of quenching of the scintillant, whereby the energy of pulses is shifted to lower energies and the efficiency of counting is compromised (*Figure 2.17*). This may be due to:

(i)   the presence of pigment which will absorb emitted light (color quenching);
(ii)   a damaged or dirty surface of the counting vial (optical quenching);
(iii)   the presence of substances which chemically interfere with the transfer of energy from the solvent to the scintillants (chemical quenching).

This potential problem requires very careful standardization of the counting procedure involving repeated checks on counting efficiency. At its simplest, this can be achieved by counting each sample before and after adding a known amount of a standard solution containing the same isotope. The efficiency with which the standard is counted will also be that of the counting of the sample. Although simple this method is somewhat tedious and also leads to contamination of the sample with other radioactive material. Alternatively, a calibration curve of counting efficiency against extent of quenching for a given isotope and a given scintillation cocktail is prepared prior to counting the sample(s).

This approach is most reliable if counting occurs over two ranges (channels) of energy of pulses by adjusting the discriminator settings

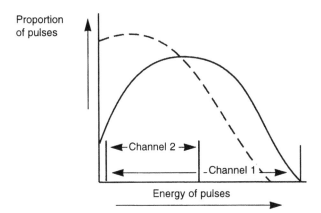

FIGURE 2.17:   *The effect of quenching on the distribution of pulses between channel 2 and channel 1. Distribution without quenching (———); with quenching (----).*

of the instrument. It depends on the fact that quenching reduces the energy of some pulses of light such that a higher proportion of the pulses are of low energy (*Figure 2.17*). If one channel is arranged to cover, say, the pulses with energy up to one-third of the $E_{\beta max}$ and the other covers the whole range, the ratio of readings in each channel will reflect the extent of quenching. This is known as the channels ratio method. In practice, this involves counting a known amount of standard radioisotope in both channels. A small amount of quenching agent is added and the counting in both channels is repeated. This procedure is repeated many times over a wide range of quenching and the ratio of counts in the two channels is plotted against the percentage counting efficiency in the broader channel (*Figure 2.18*). The calibration data obtained can usually be entered into a computer program which can then automatically be applied to correct counts of samples. Each sample is counted in both channels and the channels ratio is used automatically to indicate counting efficiency for that sample.

Both of these calibration procedures can also be conducted using an external standard of γ-emitter placed close to the counting vial in the counter. The γ-radiations cause the solvent and scintillants to generate pulses of light which are sensitive to quenching in the same way as internally generated β-radiation. Each sample is counted in a channel set to measure β-radiation-generated pulses and in another channel set to measure external standard-generated pulses. The quench factor determined for the latter readings can be applied automatically to the former readings. The external standard approach

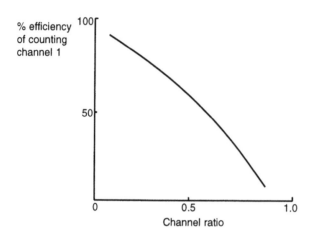

FIGURE 2.18: *A quenching calibration curve. Relationship between the efficiency of counting in a predefined channel (channel 1, Figure 2.17) and the ratio of counts in predefined channels 2:1. An increase in quenching causes the ratio to increase and the efficiency of counting to fall. Measurement of the ratio for a particular sample and application of a calibration curve allows the decrease in efficiency to be assessed and corrected for.*

can also be combined with the channel ratios method to give an external standard ratio. In addition to these standard quench procedures, some instrument manufacturers have devised more sophisticated approaches to quench correction which are reputed to be more accurate.

LSC is also usually the basis of flow-through radiodetectors that can be used on the end of LC or HPLC columns [see also the section on high-performance (high-pressure) liquid chromatography (HPLC)]. In this case the pumped eluent from the column is mixed with pumped scintillation cocktail at a predetermined ratio, prior to passing in front of the PM tube of the radiodetector. This inevitably increases the volume of peaks with some loss of resolution. It also restricts the eluting solvents to those that do not quench. In some flow-through radiodetectors the column eluent is passed over small particles of an insoluble scintillant such as calcium fluoride in front of the PM tube. These have the advantage of the eluted compounds not being diluted or contaminated. However, cell volumes are relatively large, leading to some loss of resolution. Often the efficiency of counting is less with solid rather than liquid scintillants and there is sometimes a problem of increasing background due to irreversible adsorption of traces of radioactive material on the solid particles.

The decay energy of β-radiation of ^{32}P is sufficiently high to cause water to emit a pale blue light, often referred to as the Cerenkov effect. Since the photomultiplier tubes of commercial scintillation counters are sensitive to this light and chemical quenching is not a problem, Cerenkov counting is a simple, cheap form of radioassay of ^{32}P and has been used for [^{32}P]phospholipids.

2.4.4 Gas ionization detection

The passage of β-radiation through a gas causes ionization which can be readily detected by changes in the current flowing between two electrodes. Under appropriate conditions, including a suitably high voltage across the electrodes, the rapidly moving electrons produced during ionization collide with molecules of gas (usually argon) to produce further electrons until all molecules are ionized. This amplification effect results in a very sensitive detection device, the Geiger–Müller (GM) tube. To avoid a continuous discharge, a quenching gas (e.g. a halogen or ethanol) is used, which reduces the energy of the ions. The size of the response is the same for all β-emitting radioisotopes of whatever energy since each pulse produces complete gas ionization. The instrument records the number of pulses which is directly proportional to the amount of radioisotope present. A lower voltage may be selected which results in the energy of the response pulse being directly proportional to the energy of β-radiation. This gives rise to 'proportional counting'. However, for lipid work this technique has been largely superseded by LSC and will not be considered further.

The most common GM tubes have an end window (*Figure 2.19*) through which the radiation enters the tube. The presence of a thick glass window limits use to ^{32}P whereas a thinner mica or mylar window allows β-radiation from ^{14}C, ^{35}S and ^{32}P to be measured, albeit with relatively low efficiencies (often <5% for ^{14}C). A windowless tube through which a gas mixture (often butane–helium, 1:49, v/v) flows, has to be used to measure [3]H.

GM tubes are often used to scan paper and thin layer chromatograms for distribution of radioactivity. Limitations of the geometry of the arrangements lead to poor and variable efficiencies of measurement which make accurate radioassay difficult. However, they are excellent detectors and when linked to a recorder they indicate the position and relative intensities of radioactive areas.

The position of radioactivity on chromatograms may also be determined using a windowless proportional detector. This is

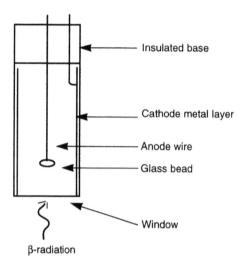

Insulated base

Cathode metal layer

Anode wire

Glass bead

Window

β-radiation

FIGURE 2.19: *Diagrammatic representation of an end window GM tube. The β-radiation causes molecules of gas, usually argon, to emit electrons and become ionized. This generates a potential difference between the anode wire and the cathode metal layer which can be measured and amplified.*

particularly useful for two-dimensional chromatograms. In the detector the anode wire runs the length of and close to the chromatogram. Areas of radioactivity cause local ionization of the gas (argon/methane) in the detector immediately above. This generates pulses in the wire, the position and frequency of which are assessed in the detector and displayed on a monitor. When sufficient pulses have accumulated to give a clear pattern of radioactivity (usually a matter of minutes) a hard copy can be made and the data recorded.

2.4.5 Gamma counting

The γ-emitters ^{125}I and ^{131}I are best assayed using a gamma scintillation counter. In this instrument a solid scintillant (usually sodium iodide) is stimulated by a pulse of γ-radiation to give a pulse of light which generates an electrical pulse in a PM tube. The size of this pulse is dependent upon the nature of the radioisotope and the frequency varies directly with the frequency of γ-radiation, i.e. the amount of radioisotope. The energy of γ-radiation from a particular radioisotope is monoenergetic unlike the β-emitters which give off β-radiation with a wide range of energies. Gamma-counting is clearly more simple than LSC of β-radiation in that, for example, no scintillant is required and no quenching occurs.

2.4.6 Radioisotopes in lipid studies

Radioisotopes feature in several types of lipid studies. These include:

(i)   metabolic studies on the conversion of a radioactive compound into some other radioactive compound(s);

(ii)   metabolic studies in which the incorporation of radioactivity from one or more precursors assists identification of a lipid product;

(iii)   studies of biological absorption and transport of lipids;

(iv)   studies of turnover of a lipid in a biological system;

(v)   microassays of a lipid by determining the dilution of specific radioactivity of a sample of pure radioactive lipid added to the system, i.e. isotope dilution analysis;

(vi)   microassays of a lipid by forming a radioactive derivative;

(vii)   as an internal standard in assays to correct for incomplete recovery of a lipid.

The availability of radioactive substances of high specific radioactivity (see below) and of extremely sensitive methods of detection and assay are essential features of all these applications. In addition, several applications depend on the fact that enzymes and transport systems rarely distinguish between molecules of the same compound that contain natural isotopes and those that contain radioactive isotopes. The main exception to this is the replacement of ^1H by ^3H which in enzyme catalyzed transfer of the hydrogen (tritium) gives rise to a significant isotope effect.

The current standard unit of radioactivity is the becquerel (Bq) which is the quantity of radioactive material that gives rise to one disintegration (release of one pulse of radiation) per second (dps). The earlier standard unit, the Curie (Ci), corresponds to 3.7×10^{10} Bq. Most commercially available radioactive substances are a mixture of radioactive and nonradioactive molecules. In multi-atom molecules, such as organic compounds, the radioactive molecules often contain only one radioactive atom (say ^{14}C or ^3H) in a specific position in the molecule. This dilution of radioactive atoms with nonradioactive atoms makes small amounts of radioactivity easier to handle and renders the substance less likely to be subject to radio-degeneration. Nevertheless, commercially available radioisotopically labeled compounds have a high specific radioactivity (Bq mol^{-1}) requiring some to be in solution in a solvent that absorbs some of the radiation energy and minimizes radio-damage (e.g. toluene, 2% ethanol). If the proposed use of the radioactive material allows, it is also wise to further dilute the radioactive compound with the same but nonradioactive compounds.

The high sensitivity of radiochemical methods requires that radiochemicals are not contaminated with other radioactive compounds. It is wise always to check the purity of radiochemicals, both at the time of purchase and after storage. The data sheet from the supplier of the radiochemical will indicate the criteria of purity used in production. This is a good guide to the best method to check purity. It will often involve a chromatographic method which can also become the basis of purification if this proves necessary.

Several companies produce ^{14}C-labeled organic compounds at a specific radioactivity of the order of 2 TBq mol^{-1} (10^{12} Bq mol^{-1}). If the strategy of an experiment requires that 10^3 Bq of radioactivity be available in a sample for accurate assay, a requirement well within the capability of counting equipment available, the assay can usually be used to measure 1 nmol or less. In several studies, dilution of the administered radioactive compound or its derivatives by the same endogenous unlabeled compound(s) may reduce this sensitivity considerably.

There are numerous examples of the applications listed in (i)–(vii) previously. Only a few pertinent points will be raised here. An example of a type (ii) study is to use the presence of ^{32}P in a lipid-soluble compound and its removal as inorganic phosphate by a phosphatase in the identification of the compound as a lipid phosphate. Similarly, the incorporation of [^{14}C]- or [^{3}H]galactose into a lipid-soluble compound and its recovery as [^{14}C]- or [^{3}H]galactose after complete hydrolysis can confirm the compound to be a glycolipid. Incorporation of radioactive sugar and phosphate into a lipid chromatographing as a single compound and their release as radioactive sugar and inorganic phosphate upon mild acid hydrolysis has contributed to the identification of prenyl phosphate sugars in glycoprotein biosynthetic studies.

In type (iv) studies, a short 'pulse' (i.e. administration of radioactivity over a short period of time) of radioactive lipid or a precursor to the lipid is added to a biological system. After sufficient time, during which the exogenous radioactive material equilibrates with the endogenous unlabeled lipid in question, the system is sampled for the radioactive lipid over a period or time. The rate of disappearance of radioactivity associated with the lipid allows the half-life of the lipid to be calculated. Sometimes the addition of a radioactive precursor to a system is preferred because this is more likely to ensure that the radioactive lipid formed is in the natural site within the system. In this case the 'pulse' of radioactive precursor is often followed by a 'wash' or 'chase' of a large excess of nonradioactive precursor. This effectively stops further significant incorporation of radioactivity into the lipid and makes the timing of the 'pulse' more accurate. It may occasionally also influence the flux of metabolites through the system, so causing a change of turnover depending on the natural control of synthesis and degradation of the lipid.

Radioisotope dilution analysis is a term sometimes used for the application of type (v) studies. It is particularly useful for the assay of lipids which have to be subjected to purification procedures with attendant heavy losses before reliable assay can occur. In this case a known amount (A mol) of a radioactive sample of the pure lipid of known specific radioactivity (B Bq mol^{-1}) is added to the system. After equilibration with the endogenous lipid a sample of pure lipid is prepared and its specific radioactivity (C Bq mol^{-1}) is determined. It follows that the original amount of endogenous lipid present was ($A \times B/C) - A$ mol.

2.4.7 RIA and related techniques

Substoichiometric isotope dilution analysis (SIDA) is a valuable extension of isotope dilution analysis and avoids the measurement of the mass of purified lipid recovered from the sample. This makes a very sensitive assay, which when combined with the specificity and sensitivity of immunochemistry, gives rise to the powerful technique of RIA. In RIA an antibody specific to the lipid in question as antigen (or more accurately as hapten since most lipids are nonimmunogenic unless first complexed with a protein) is used as a reagent. At its simplest a known amount (a mol) of pure radioactive lipid antigen of specific radioactivity A Bq mol^{-1} is mixed with a substoichiometric quantity (often 50–70%) of antibody (b mol equivalent). The antibody–antigen complex is recovered separately and assayed for radioactivity R_1. The procedure is repeated but the sample to be assessed is also present (see *Figure 2.20*). If the amount of radioactivity recovered in the presence of y mol of lipid is R_2 Bq the following relationships hold:

$$R_1 = kb \times A \qquad \text{(i)}$$
$$R_2 = kb \times \left(\frac{a}{a+y}\right) A \qquad \text{(ii)}$$

in which k is a constant that takes account of the avidity of antibody–antigen binding.

Substituting (i) in (ii):

$$R_2 = R_1\left(\frac{a}{a+y}\right) \text{ or } y = a\left(\frac{R_1}{R_2} - 1\right).$$

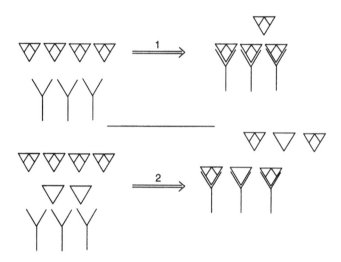

FIGURE 2.20: *Diagrammatic representation of the principle of radioimmunoassay (RIA). The sample of known amount of radioactive lipid (▽) saturates the substoichiometric amount of specific antibody (Y) in assay 1. In assay 2 the same amount of pure radioactive lipid is mixed with the sample to be assayed for the quantity of this nonradioactive lipid (▽) present. In this particular example the amount of lipid (▽) was one third of the total lipid (▽ + ▽). Comparison of the amount of radioactivity retained by the same amount of antibody (Y) in the two assays and knowledge of the amount of radioactive lipid (▽) added allows determination of the amount of nonradioactive lipid (▽) present.*

In this way only the mass of radioactive standard and the two radioactivity assays are required to calculate the mass of lipid present. Often a calibration curve is prepared using known quantities of pure nonradioactive lipid in place of the sample to be assessed, as in *Figure 2.21*. In this case one radioassay (R_2) and consultation of the calibration curve is then sufficient to give the mass of the lipid antigen present in the sample.

An important development of RIA has been immunoradiometric assay in which the antibody and not the antigen is radioactive, usually due to [125I] (see also Section 2.5.4). In one version of this assay a stoichiometric excess of selected high avidity nonprecipitatory antibodies is used, ensuring that essentially all lipid antigen is taken up to form a soluble antigen–[125I]antibody complex. The remaining uncomplexed [125I]antibody is then allowed to react with a sample of pure antigen that is coupled to an insoluble matrix. This radioactive, insoluble complex is separated by centrifugation and the supernatant, still containing the soluble sample antigen–[125I]antibody complex, is

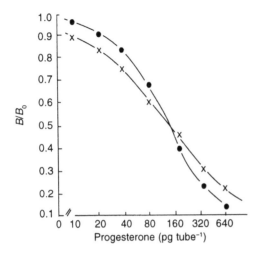

FIGURE 2.21: *Calibration curves for the RIA of progesterone using [³H]-progesterone (x-x-x) or a [¹²⁵I]glucuronide of progesterone (●-●-●), and an antiserum raised against 11α-hemisuccinate progesterone–bovine serum albumin conjugate. B/B₀ is the ratio of radioactivity (B) recovered in the presence of nonradioactive progesterone in the assay to that (B₀) recovered in its absence (i.e. assay 2 : assay 1 in Figure 2.20). Redrawn from ref. 15 with permission from Blackwell Science Ltd. Data from ref. 16.*

assayed for radioactivity. Calibration of the [¹²⁵I]antibody with known masses of standard lipid antigen allows direct calculation of the mass of lipid in the sample. The availability of monoclonal antibodies for a variety of lipids has greatly enhanced the popularity of immuno-radiometric assay.

2.5 Immunochemical methods

2.5.1 Antibodies to lipids

It has been suggested that the high selectivity and sensitivity of antibody–antigen interactions, coupled with increasingly sophisticated methods of raising large quantities of antibodies to order and of using them, will in future allow immunochemical methods to rival many physical methods for elucidation of structure and microassay of biochemicals. This futuristic view is an overstatement with regard to lipids for they are in general not very immunogenic. However the glycolipids present a special case since antibodies of high avidity and specificity are available for most of them. Interestingly, naturally occurring human monoclonal antibodies to blood group determinants such as for groups A, B and H can be isolated from the blood serum of individuals. Since these specific determinants on the erythrocyte cell surface are glycolipids there is a potential supply of this immunochemical reagent for some glycolipids. This situation highlights the high specificity of these antibodies because blood group A and B determinants differ only in the replacement of N-acetylgalactosamine in the former by galactose in the latter (i.e. replacing an amino-acetyl group by a hydroxyl group), as part of an otherwise identical oligosaccharide of six to ten or more monosaccharide residues (depending on the particular glycolipid) linked to ceramide (see Section 8.2.2).

Pure glycolipids, like other lipids, are not very immunogenic. To stimulate the production of specific antibodies in an animal these lipids need to be administered as a conjugate either covalently linked to a foreign protein (as a hapten) or as part of the bilayer of a liposome. Even then, while the immunogenic activity of these complexed lipids is high in rabbits it is often much lower in mice, prejudicing the generation of sustained monoclonal antibodies by cell fusion technology. Because lipid molecules are relatively small it is likely that the antibodies produced to these in animals stimulated by lipid conjugates recognize only one or a small number of epitopes and are therefore in fact monoclonal products or at most from only a small number of clones.

Immunochemical methods have been developed for the assay of quantitatively major lipids such as phospholipids and triglycerides. However, the group of lipids where detection and assay has benefitted particularly from immunochemical methods is the steroid hormones. In this case conjugates, usually of hormone with serum albumin, are sufficiently immunogenic to stimulate generation of antibodies with high avidity. Details of the methodology for the production and use of such antibodies is beyond the scope of this book and can be found elsewhere (see e.g. ref. 15). However, some lipid analytical procedures in which they are applied are outlined below (see also refs 17 and 18).

2.5.2 Hemagglutination in detection and assay

One of the earliest applications of immunochemical methods to glycolipid analysis involved the inhibition of hemagglutination. This method relies on divalent antibodies to a particular glycolipid, X, cross-linking between single molecules of X in adjacent erythrocytes (*Figure 2.22*). If several molecules of X exist in each erythrocyte this leads to aggregation of erythrocytes (hemagglutination). This is

(a)

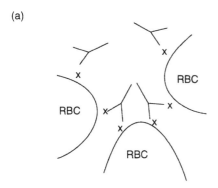

(b)

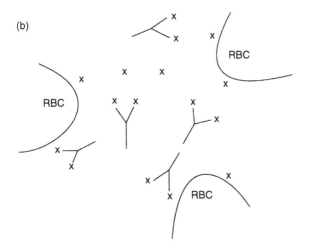

FIGURE 2.22: *Diagrammatic representation of antibody-induced agglutination of red blood cells (RBCs). (**a**) Hemagglutination occurs through the anti-X cross-linking membrane-bound molecules of lipid X in different RBCs. (**b**) In the presence of lipid X in an added sample (preferably presented at the surface of a liposome) fewer antibody molecules remain for cross-linking RBCs and hemagglutination is reduced. Appropriate calibration with known amounts of X provides a convenient assay for X.*

usually carried out in a microtiter plate and agglutination is recognized as a 'red carpet' in a plate well. A glycolipid preparation will inhibit this agglutination if it contains the antigen X by competing for binding sites of the antibody (*Figure 2.22*). Calibration of the inhibitory effect by using pure antigen X allows an assessment of the amount of X in the glycolipid preparation. The sensitivity of the method is enhanced by presenting the inhibiting glycolipid (preparation and standard) in a liposome suspension of phosphatidyl choline and cholesterol (see e.g. Section 2.6). For this procedure approximately 10 µg of the glycolipid X will be required.

If erythrocytes do not normally contain the glycolipid of interest, the pure glycolipid can readily be incorporated into the erythrocyte plasma membrane before the agglutination studies are initiated.

2.5.3 Immunostaining and TLC

For several years, the immunostaining of TLC chromatograms for the detection and assay of glycolipids has proved popular. The chromatogram is treated with a radiolabeled specific antibody (usually with [125I]) to stain only the glycolipid antigen even in the presence of other overlapping glycolipids (*Figure 2.23*). In this method, flexible TLC plates (plastic- or aluminum-backed) are used. After chromatography the dried chromatogram is stabilized by dipping in a solution of poly(isobutyl methacrylate) (0.5% w/v) in diethyl ether and air drying. The treated chromatogram is then wetted by immersion in phosphate-buffered saline (pH 7.2) (PBS) containing bovine serum albumin (2% w/v) – buffer A. It is then placed horizontally in a shallow dish and overlayed with a minimum volume of a solution of labeled antibody in buffer A. After a few hours the chromatogram is washed several times with PBS and allowed to dry before autoradiography. Detection of the 125I may be by auto-radiography (see Section 2.4.2). The chromatographic mobility and antibody staining of the glycolipid serve to identify it. If calibrated with known amounts of pure glycolipid antigen, preferably added to a portion of the sample before TLC, the method can be made quantitative either by scanning with a radiodetector or by removing the radioactive glycolipid–antibody complex plus TLC adsorbent and assaying for 125I in a gamma counter. Modifications of the procedure which improve sensitivity involve using an unlabeled mouse anti-glycolipid immunoglobulin G (IgG) as first antibody and staining the resultant complex either with a second anti-mouse IgG [125I]antibody or [125I]protein A. This latter protein is produced by *Staphylococcus aureus* and binds to mouse IgG strongly and specifically. This method has also been adapted to demonstrating the binding to glycolipids of other radioactive materials (e.g. lectins, some toxins, some bacteria, some viruses).

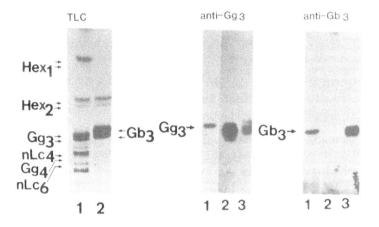

FIGURE 2.23: *The use of ^{125}I-labeled specific antibodies to the glycolipids gangliotriaose (Gg₃) and globotriaose (Gb₃) (see Table 8.2) in immunostaining of TLC plates. The figure compares the glycolipids of undifferentiated (M1–) and differentiated (M1+) myelogenous leukemia cells. The left panel shows the glycolipids of M1– and M1+ (lanes 1 and 2, respectively) stained by the orcinol-sulfuric acid reagent (see Section 8.2.3). The central panel shows staining with anti-Gg₃ of standard Gg₃ (lane 1) and of the lipids of M1– (lane 2) and M1+ (lane 3) cells. The right panel shows staining with anti-Gb₃ of standard Gb₃ (lane 1) and of the lipids of M1– (lane 2) and M1+ (lane 3) cells. The absence of Gb₃ in M1– cells is clearly shown as is a decline in Gg₃ and its replacement by Gb₃ in the differentiation between M1– and M1+ cells. Reproduced from ref. 19 with permission from Blackwell Science Ltd. Data from ref. 20.*

2.5.4 Immunoradiometric assays

Most lipids bind readily to the wells of plastic multiwell microtiter plates. This enables them to be identified and measured by a solid-phase immunoradiometric assay (IRMA) if an appropriate high-binding antibody is available. This has been the basis of some methods of glycolipid detection and assay for several years. A sample of a liposome preparation containing the glycolipid in question together with phosphatidyl choline and cholesterol is adsorbed on to the surface of the wells of a plastic microtiter plate. The bound lipid is exposed to the antibody (X) and that antibody that binds is detected and assayed by binding either an anti-X [125I]antibody or [125I]protein A. Unbound material is removed at each stage by washing and the amount of ^{125}I bound to the well is determined in a gamma counter. The assay is calibrated by adding known amounts of standard glyco-lipid to the original preparation and determining the amount of ^{125}I bound. It is sensitive to 1–2 ng of glycolipid.

As mentioned in Section 2.4.7 the availability of monoconal antibodies (MAb) of high avidity has been essential for the development of good IRMA methodology, in which the antibody carries the label to be measured and is used in excess. The advantages of MAb-based IRMA can be summarized as:

(i)   shorter incubation times (2–3 h) are required than for RIA;
(ii)   better defined specificity than for RIA;
(iii) a regular supply of identical samples of antibody.

When allied to using solid phase binding as the basis of separation of reacted and unreacted antibody the following further advantages make the method very popular:

(iv)   greater sensitivity and precision;
(v)   greater robustness in inexperienced hands.

Thus solid phase MAb-based IRMA provides a powerful ultrasensitive assay procedure which may be expected to be used frequently for microassay of lipids that have functional groups, whereby they may be linked to proteins or particulate polymers in order to become immunogenic.

2.5.5 Enzyme-linked immunosorbent assays

In the most sensitive IRMA the detection of ^{125}I limits the sensitivity of the assay. If antibodies are labeled with enzymes instead of ^{125}I the limit can sometimes be reduced considerably (10^3-fold for enzyme labels that themselves have a very sensitive assay procedure). In addition to the amplification effect of the enzyme itself, advantages of enzyme immunoassays (EIA) include their relative cheapness, safety, the long shelf-life of reagents and the use of widely available, standard equipment (see e.g. ref. 18). Disadvantages include the possible interference in enzyme activity by components of the lipid fraction being assayed. Just as there are several versions of RIA and IRMA, so there are corresponding versions of EIA. Some of these are grouped under the term enzyme-linked immunosorbent assay (ELISA), in which one of the reactants is immobilized on a solid-phase matrix (see *Figure 2.24*). For ELISA of lipids the lipid is usually bound to a solid phase and the antibody is measured either by virtue of itself carrying enzyme or by using a second antibody which carries an enzyme. The latter is often preferred because if all primary antibodies are from the same animal, only one enzyme-linked antibody is required. Noncompetitive ELISA in which excess enzyme-linked antibody reacts with immobilized lipid preparation or immobilized lipid standard and the amount of binding is determined is probably the most common version for lipid assays.

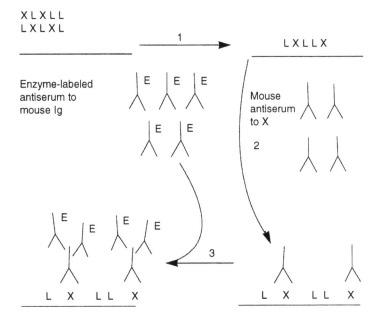

FIGURE 2.24: *Diagrammatic representation of the principles of ELISA of lipids. In step 1, a mixture of the lipid (X) to be assayed and of other lipids (L) is bound to a solid phase. Mouse monoclonal antiserum to X is applied in step 2 and binds to X. Step 3 involves addition of enzyme-labeled antibody (in this case polyclonal) to mouse Ig and several molecules become attached to each immobilized molecule of mouse antibody. Assay of the amount of enzyme bound provides a sensitive, specific measure of the amount of X present.*

2.6 Analytical methods involving enzymes

Pancreatic lipase, which hydrolyzes triacylglycerol, and various phospholipases have all proved useful in the determination of lipid structures. Some examples are given in this section.

2.6.1 Phospholipases

A number of different phospholipases are available commercially. Their specificity and nomenclature are shown in *Figure 2.25*. Naturally occurring phospholipids belong to the L-glycerol 3-phosphate series, using the Fischer projection of glycerol in which the hydroxyl at C2 appears on the left (*Figure 2.26*). As the figure shows, L-glycerol 3-phosphate is identical to D-glycerol 1-phosphate, but it is

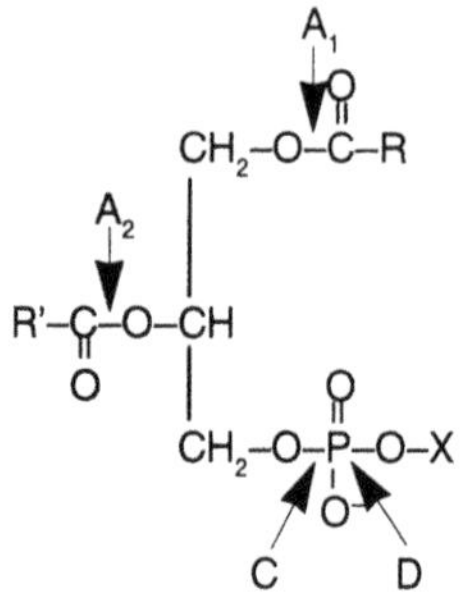

FIGURE 2.25:   *Sites of action of phospholipases A, C and D.*

now more common to use stereospecific numbering (*sn*) in which this stereoisomer is called *sn*-glycerol 3-phosphate.

Phospholipase A$_2$ is available commercially in purified form from the pancreas or from bee or snake venom. It selectively removes the unsaturated fatty acid from the 2-position of phospholipids, leaving a '*lyso*'-phospholipid, so-called because of its ability to lyse red cells. This enzyme has been used to determine the nature of the fatty acids linked to the 1- and 2-positions of phospholipids such as phosphatidylcholine. Fatty acids can be released from lysophosphatidylcholine by mild alkaline hydrolysis. GLC is used to separate and identify the fatty acids released.

Phospholipase A$_1$ (*Figure 2.25*) has been purified from the pancreas, brain and various micro-organisms, but it may not be completely specific. It can also hydrolyze 1-acyl lysophospholipids and act as a phospholipase B, removing both 1-acyl and 2-acyl residues from a phospholipid.

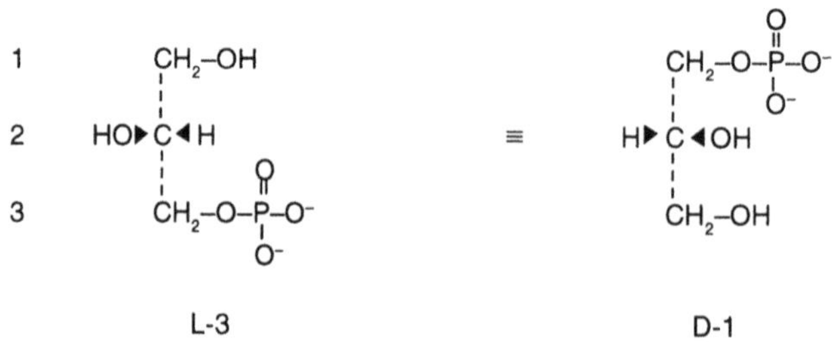

FIGURE 2.26:   *Stereospecific numbering of glycerol phosphate. The* L-3 *(≡* D-1*) compound is referred to as* sn-*glycerol 3-phosphate.*

Two types of phospholipase C are available, both obtained from micro-organisms. The enzyme from *Clostridium perfringens* hydrolyzes phosphatidylcholine, sphingomyelin and all the phospholipids except phosphoinositides. The enzyme from *Bacillus cereus*, on the other hand, hydrolyzes only phosphatidylinositol and its derivatives. Release of diacylglycerol by the action of phospholipase C confirms the structures of the phospholipids. In the case of sphingomyelin, *N*-acylsphingosine is released. The water-soluble phosphates which are the other products of phospholipase C action can be identified by paper chromatography or by their behaviour on ion exchange columns. Some plasma membrane proteins in mammals and parasitic protozoa are covalently linked to glycosylated phosphatidylinositol. The phospholipase C of *B. cereus* removes such proteins from the membrane in many cases by hydrolyzing the membrane phosphatidylinositol. Several glycosyl phosphatidylinositol-specific phospholipase C forms are also known.

Phospholipase D, available commercially from cabbage and other sources, hydrolyzes a variety of phospholipids including sphingomyelin and cardiolipin. Phosphatidic acid and choline are produced from phosphatidylcholine, both of which may be identified by chromatography. The enzyme is also capable of transphosphatidylation, in which a primary alcohol instead of water accepts the phosphatidyl group. In the presence of ethanol, for instance, phosphatidylethanol is produced. This type of reaction has proved useful in the synthesis of modified phospholipids and in the identification of phospholipase D in tissues.

2.6.2 Stereospecific analysis of triacylglycerol

Pancreatic lipase will eventually remove all three fatty acids from triacylglycerol, but the *sn*-1 and *sn*-3 acyl groups are preferentially removed in the early stages. Brockerhoff and Yurkowski [21] made use of this to identify the fatty acids in all three positions of a triacylglycerol. A slightly easier method, devised by Lands [22] is illustrated in *Figure 2.27*. It makes use of two further enzymes, diacylglycerol kinase, which uses ATP to phosphorylate only the *sn*-3 hydroxyl of glycerol and phospholipase A_2, which removes the fatty acids only from the *sn*-2 position. Saponification of the resulting 1-acyl lysophosphatidic acid releases the fatty acids from the 1-position. The 1- and 2-fatty acids can then be identified by GLC. Methanolysis of the original triacylglycerol and GLC of the resulting methyl esters gives the total fatty acids of all three positions. Comparison of this with the 1- and 2-fatty acids allows the fatty acid composition of position 3 to be calculated.

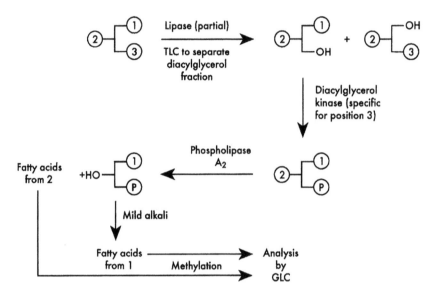

FIGURE 2.27: *Stereospecific analysis of triacylglycerol.*

2.6.3 Enzymes in quantifying lipids

Cholesterol oxidase has been used in determining cholesterol concentration in blood plasma, but chemical methods are more widely used. Similarly, lipase could be used in the determination of triacylglycerol levels, but chemical estimation of the fatty acids or glycerol released would still be necessary, so it is more convenient to hydrolyze the triacylglycerol with alkali. Furthermore, there is always the possibility that enzymes used in analysis may be inhibited by other substances in the sample.

2.7 Presentation of lipids in aqueous systems

Apart from the highly polar phospholipids and glycolipids, lipids do not dissolve in water. It may be necessary, therefore, to disperse them in an aqueous system for work with enzymes. Physiologically, this is done in the small intestine, where emulsification with bile salts acting as detergents enhances the action of pancreatic lipase on triacylglycerol. In the laboratory, an insoluble lipid such as triacylglycerol may be dispersed in an aqueous system by adding a detergent such as Tween 20 (polyoxethylenesorbitan monolaurate, 5–10 mg per 100 mg triacylglycerol) and using a sonicator. A bile salt such as sodium deoxycholate can be used in the same way as Tween 20. Another dispersing agent is the nonionic detergent Cutscum (*iso*-octylphenoxypolyethoxyethanol).

Free fatty acids and lysophospholipids readily bind to the protein albumin and are thus made water soluble. Phospholipids form bilayers in aqueous media, so that their hydrophobic acyl chains are hidden from the water. Such bilayers close up to form multi-layered vesicles. Sonication under various conditions will alter the number of bilayers in the liposomes. Phophatidylcholine is the lipid most often used for the preparation of liposomes, sometimes with the addition of 1–5% (w/w) of phosphatidic acid.

If the liposomes are prepared in an aqueous medium containing a water-soluble drug, this drug will be trapped inside the liposomes. Such liposomes are of interest in medicine as drug delivery agents which provide a slow release system within the body. Liposomes are also used in moisturizing creams for the skin.

2.8 Care and safety during the analysis of lipids in biological materials

2.8.1 General considerations

The use of special solvents and chemicals as well as equipment for the analysis of lipids is an important feature of the techniques discussed in this chapter. If properly used they present a minimal risk to health. If improperly used the risk of the potential hazards being realized becomes unacceptably high.

In the UK, The Health and Safety At Work Act (1974) makes provision for the health, safety and welfare of all people at work and for the controlled use of dangerous substances and emissions. The Act places upon everyone a responsibility for care in both specialized and ordinary activities. Its success depends heavily on the individual wishing to work safely.

When preparing to carry out experiments involving lipids of biological origin it is essential that experimenters are fully aware of the potential hazards and of the current regulations and laws pertaining to those activities in their place of work. It is incumbent upon each individual laboratory worker and supervisor to assess and minimize the risks of their experiments both to themselves and to others. In many countries this is a legal obligation.

In the lipid laboratory hazards of four types may need particular attention. These concern: (i) the use of chemicals and solvents that may be hazardous to health; (ii) increased risk of fire due to handling flammable solvents; (iii) the use of radioisotopes with their own particular hazards; and (iv) the biological hazards associated with the use of pathogenic organisms or of tissues or body fluids that may harbour potential pathogenic organisms. These are discussed in the following pages in the context of regulations and laws existing in the UK. Although the risk of accident and of personal injury cannot be eliminated completely, it should be stressed that proper assessment of hazards and preparation of appropriate protocols in the light of national/local recommendations and regulations can reduce the risk to acceptable levels.

2.8.2 Chemical hazards

Work in the UK with substances which are hazardous to health is subject to the requirements of the Control of Substances Hazardous to Health (COSHH) regulations of 1994. This is an updated version of the 1988 regulations which came into force in January 1995. The main changes relate to enforcing directives relating to biological agents, leaving the control of chemical agents as in the 1988 regulations (see e.g. ref. 23). The substances concerned include those which are categorized as very toxic, toxic, harmful, corrosive or irritant. With regard to chemicals obtained commercially, the category of hazard is indicated on the label of the container (*Figure 2.28*). The UK Carcinogenic Substances Regulations (1967) are still in operation. Some carcinogens are prohibited under these regulations and are not available commercially. Others are listed as Controlled Substances and are available for use under carefully controlled conditions.

The health risks of all chemicals to be used in a laboratory have to be assessed and recorded bearing in mind the specific, possibly unique situation, in which the use will occur. This will involve consideration of the toxicity of the substance, the quantity involved, its physical form and the route by which exposure to the hazard may occur. If it is considered that a serious risk to health is involved, an alternative substance should be sought. If this is not available control measures have to be planned and recorded. This may involve, for example, working in a fume cupboard, using a solution rather than a powder, containment in a closed vessel or wearing particular protective clothing. The record needs to detail also the procedure for dealing with spillages and, for very toxic materials, the availability of antidote. In the case of hazards that accrue from repeated exposure to a substance, an exposure limit is set and a method of ensuring that this is not exceeded will need to be established. The records of risk assessment must be available for inspection by officials on request.

The proper handling of some hazardous substances may require special training of the individual concerned. Hazard data sheets for each chemical sold are provided on request by the manu-facturer/supplier. Several reference books (see e.g. ref. 24) and data handbooks (see e.g. refs 25–27) are also available.

Suppliers of chemicals are also under an obligation to provide guidance towards the safe handling of their products. For example, in Europe most companies use symbols and phrases approved by the EC for classification, packaging and labeling. These are used on labels on containers of chemicals and can be found in products catalogs. Several companies provide an extensive explanation of the symbols or pictograms (see e.g. *Figure 2.28*) in the catalog. This may also contain an explanation of the number code on the label that refers to specific EC approved risk and safety phrases. Some companies provide explicit interpretation of the hazard symbol and advice in case of accident on the container label. Others give the class number of the United Nations Hazard Classification (*Table 2.8*). Some catalogs also provide advice on first aid, fire extinction, spillage treatment and waste disposal for their products, which may be coded by letter or number on the label of the container and in the catalog entry for the product.

Storage and use of all chemicals must minimize the possibility of accident. Hazardous chemicals should never be stored: (i) in large quantities; (ii) in a dangerous manner; (iii) above eye level; and (iv) without adequate labeling. Toxic, flammable or corrosive materials should be stored at low levels in suitable cupboards or bins and away from sources of heat.

All chemicals should be handled carefully. On the bench they should not be placed in a precarious position or left uncapped. They should be returned to store as soon as possible and a cluttered bench should be avoided. Minor spillages should be cleared up immediately as indicated on the risk assessment sheet. Some operations are best carried out in a fume cupboard (offering good ventilation and a physical barrier) which should be seen as an extension of the bench for some hazardous work and not as a storage area.

The cleaning up of major spillages or spillages of particularly toxic material may require special assistance. The logistics and mechanism of this should have been anticipated in the risk assessment sheet.

FIGURE 2.28: Hazard symbols (black on an orange or yellow background) in common use by suppliers of chemicals and displayed in laboratories. These are used as follows:

Toxic: *substances which present a serious risk of acute or chronic poisoning by inhalation, ingestion or skin absorption. Those substances that are toxic at low concentration are termed* very toxic. *Those that are included in Schedule 1 list of the Poisons Rules (1978) are sometimes identified by the appearance of S.1 next to the symbol.*

Corrosive: *substances which destroy living tissue on contact.*

Harmful: *substances which present a moderate risk to health by inhalation, ingestion or skin absorption. Often these noxious materials are indicated by Xn to distinguish them from irritant substances. Those, which although not corrosive, are liable to cause inflammation through contact with the skin or mucous membranes, are identified by Xi and the word* irritant.

Radioactive: *substances which emit α-, β- or γ-irradiation.*

Radiation: *substances which are a source of potentially damaging radiation (the 'radioactive' and 'radiation' symbols are often used synonymously).*

Flammable: *liquids having a flash point of 21°C or more and below or equal to 55°. Highly flammable substances include: (i) liquids having a flash point of below 21°C; (ii) those which in contact with water or damp air evolve highly flammable gases; (iii) gases which are flammable in air at normal pressure; (iv) those which may become hot when in contact with air; and (v) those which may readily catch fire after brief contact with a source of ignition and which continue to burn afterwards. Extremely inflammable liquids are those with a flash point below 0°C and a boiling point of or below 35°C.*

TABLE 2.8:   *United Nations Hazard Classification of Chemicals*

Class number	Hazard
1	Explosive
2	Gases
3.1	Flammable liquids: flash point below −18°C
3.2	Flammable liquids: flash point −18 to +23°C
3.3	Flammable liquids: flash point +23 to +61°C
4.1	Flammable solids
4.2	Spontaneously combustible
4.3	Dangerous when wet
5.1	Oxidizing agent
5.2	Organic peroxides
6.1	Poisonous
7	Radioactive
8	Corrosive
9	Miscellaneous dangerous substances
NR	Nonregulated

Small quantities of many substances that are soluble in or miscible with water may be washed down the sink with plenty of water. Some substances, e.g. strong acids and alkalis, should be neutralized first. Organic solvents need to be stored in metal drums prior to removal and disposal by trained professionals. All drums should carry an accurate description of their contents and a more detailed record should be kept of the additions to the drum. Chlorinated hydro-carbons should be stored separately from other solvents.

2.8.3 Fire hazards

The majority of organic materials are flammable and several volatile solvents present a particular fire hazard to lipid researchers. The volumes of highly flammable liquids in a work area must be kept as small as is reasonably practical having regard to the operations being carried out. Volumes stored in fire-resistant cupboards or bins for immediate use should be limited (50 l is reasonable). The containers and the bins/cupboards must be properly labeled. Particular care should be taken not to use flammable, volatile liquids in the vicinity

---

Oxidizing: *substances which produce highly exothermic oxidizing reactions in contact with other substances.*
Explosive: *substances which may explode under the effect of heat or which are more sensitive to shock or friction than is dinitrobenzene.*
Biological hazard: *the risk of infection from pathogenic micro-organisms known to be present or potentially present in biological material. This includes the potentially harmful results of genetic manipulation, especially of micro-organisms known to be pathogenic or resistant to antibodies.*

of a source of ignition such as a gas burner, hot surface, electric motor, light switch or bimetallic strip on/off device.

In the event of a solvent fire, water should not be used to fight it for many solvents float on water and this simply spreads the fire. Smothering with a fire blanket is often the best way to deal with a small solvent fire. Fire extinguishers are best used to back up the smothering. Major fires should be left to trained fire fighters who should be called immediately. In a fire the location of gas cylinders, which become potential bombs, of organic solvents and of any special hazards (for example radiochemicals or highly toxic materials) should be indicated to the fire fighters.

2.8.4 Radioisotope hazards

The use of radioactive substances and of apparatus producing ionizing radiations is governed in Great Britain by the Ionizing Radiation Regulations (1985). The acquisition and disposal of radioactive materials is regulated by the Radioactive Substances Act (1993), compliance with which is monitored by HM Inspectorate of Pollution (HMIP). Institutions using radioisotopes have a framework of control and monitoring of use of radioisotopes which involves specific responsibilities from the individual worker through departmental Radiation Protection Supervisors to institutional Radiation Protection Officers (RPO).

The acquisition and disposal of radiochemicals by any individual must comply with conditions and limits agreed with the RPO working within limits imposed by HMIP. This requires the individual to present a brief outline of the type of work to be undertaken to the RPO. Approval involves the formal registration of the individual and the project. Usually work will be allowed only in registered laboratories that have been inspected and approved for safe usage of radioisotopes.

The law states that: 'Every individual worker with radioactive substances has a duty to protect himself and others from any hazard arising from his work. He must not expose himself or others to a greater extent than is necessary for the purpose of his work'. To this end, it is mandatory to maintain adequate records of purchase, storage, use and disposal of radioisotopes; all radioactive substances and rooms for their use should be properly labeled (*Figure 2.28*). Film badges, sometimes finger badges, should be worn to monitor exposure to radiation. Laboratory coats and gloves should always be worn. All radioactive work is best carried out on stainless steel trays to contain splashes. Bench tops are best protected by the use of Benchkote. The individual should be protected from radiation by a transparent shield of Perspex for ^{32}P and containing lead for ^{125}I. High levels of radio-activity ($>10^6$ Bq) should be handled and stored only in a well equipped laboratory specially designed for this. Equipment, glassware, protective clothing, etc. should not be moved out of this room to other laboratories for fear of contamination. All equipment and work areas should be monitored with a Geiger counter and cleared to within the local acceptable limits after a radioisotope experiment. If clothing is contaminated it should be disposed of as radioactive waste.

Radioactive waste, within agreed disposal limits, may be poured down laboratory sinks designated for this purpose, with copious amounts of water. Escape of radioactive gas (e.g. of $^3H_2$) via fume hoods is allowed within agreed limits. Low level contamination (e.g. −0.4 MBq per bin) of paper can usually be disposed of by incineration. Other radioactive waste is usually required to be placed in special plastic bags or, if liquid, in container drums which are labeled obviously as radioactive with an estimate of the amount and nature of radioisotope concerned, prior to its removal by a registered waste disposal company.

2.8.5 Biological hazards

Biological materials are a potential source of hazards including infection, sensitization and environmental damage (see e.g. ref. 24). The COSHH (1994) regulations provide the main framework for control of these hazards. The lipid analyst of biological materials should be aware of these hazards and assess the risk of exposure (to the analyst and to others). Any risk should be controlled by a mixture of procedural and engineering means.

In the UK, advice on the handling of pathogenic organisms is available in the report of the Advisory Committee on Dangerous Pathogens (ACDP), 'Categorization of Pathogens According to Hazard and Categories of Containment' (second edition, 1990), which groups micro-organisms into four broad hazard categories. The corresponding containment level describes the minimal laboratory conditions under which a particular pathogen is permitted to be handled, taking account of the organism's pathogenicity, route of transmission, epidemiological consequences and host susceptibility. Vaccination is seen as a containment feature when working with some organisms. Training is viewed as an essential aspect of safe working with pathogens, taking particular account of special procedures and equipment needed to minimize the risks.

Generally it is wise to wear the correct protective clothing and to use a safety cabinet or hood which filters the air and has an air flow away from the operator if there is the slightest risk of pathogenicity. The ACDP report makes these precautions mandatory.

Human tissues, cells and body fluids should be treated as potential sources of pathogenic organisms. Particular concern attaches to the potential presence of Hepatitis B virus and human immunodeficiency virus (HIV). Immunization against Hepatitis B is strongly recommended. Advice on containment arrangements needed to handle HIV, whether known to be or potentially present, is given in the ACDP report entitled 'HIV – The Causative Agent of AIDS Related Conditions' (second revision of Guidelines, January 1990).

The use of genetically manipulated material is regulated by the Genetically Modified Organisms (Contained Use) Regulations (1992). A separate relevant publication is 'Genetically Modified Organisms (Deliberate Release) Regulations' (1992). Any use of this material is controlled by a local Biological Safety Officer and Committee. It is assessed by relevance to a published approved protocol (Health and Safety Executive (HSE), Advisory Committee on Genetic Manipulation (AGGM), Department of the Environment (DOE) note of 7 September 1993) and the proposed work must be judged as acceptable by the Biological Safety Officer before it can commence.

2.8.6 Other hazards

Particular care should be taken with the use of sharp instruments such as needles and scalpels. If disposable after use they should be placed in a special container. The use of glassware has its attendant risks. In particular the collapse of glass chromatography columns, or of pipettes when applying pressure from one end, is a notorious source of badly cut hands. Great care is needed. Excessive or uneven pressure should certainly be avoided. Not only are cuts themselves to be avoided but it should be realized that breaking the skin with contaminated glass or sharps risks toxicity/infection.

Compressed gas cylinders, although quite safe it used properly, present hazards of several sorts if used improperly. They are usually heavy and cumbersome and require special trolleys for their transport to and from and around the laboratory, and also special fixing arrangements at their position of use. Pressure gauges should be attached properly and the valves opened cautiously. If possible always test the cylinder valve in a well ventilated area before taking it into the laboratory. The sudden excessive opening of stiff valves is a potential source of serious accident.

Much of the equipment used in lipid analysis is powered by electricity and in the UK its use must conform with the Electricity at Work Regulations (1989). The fundamental hazard is one of electrocution and all precautions are aimed at reducing this. The equipment and wiring must remain electrically safe. It should be inspected and tested and if necessary serviced regularly by a qualified electrician. Loose screws, wires, potential shorting, sparking or overloading are also a major fire risk in the presence of organic solvents.

2.8.7 Personal preparedness and protection

If an accident does occur the consequent damage to persons and/or building and contents can often be minimized if the individual is well prepared. It is essential to be aware of the positioning of safety equipment and materials. The laboratory worker must ensure that he/she knows the location of fire extinguishers and fire blankets, of first aid boxes and of the name and location of the nearest laboratory fire officer and first aider. The individual should be familiar with the location and handling of antidotes to any poisons that are to be used. He/she should also know the local procedure for dealing with noxious vapors.

Clean laboratory coats made of an absorbent material such as cotton offer protection from chemical splashes and from infectious material. They should not be worn outside the laboratory, especially in communal rooms such as common rooms, cafeteria and libraries, where transfer of hazards on the coat extends risks unnecessarily. Feet are best protected by wearing sturdy shoes, avoiding especially open-toed shoes or sandals. In general, clothing liable to generate static electricity (e.g. nylon) should be avoided.

Some activities, for example shaking fluids and applying fluids under pressure, risk damage to the eyes. Goggles or safety spectacles should be worn on these occasions and it is sensible to develop the habit of always wearing them in the laboratory.

Lightweight disposable gloves are recommended to protect the skin from infectious materials, from dry powders (for example when weighing them) and from splashes. Heavy duty gloves are essential when handling corrosive substances such as strong acids or alkalis. Some relatively innocuous solvents require cautious handling for they may remove lipids from the skin, so reducing the permeability barrier to infection and to some toxic materials. Other solvents although not presenting a major acute toxic risk are recognized as chronically damaging to the environment and should be used in minimum quantities. The availability of ozone-depleting halogenohydrocarbons is being phased out under international agreements. Under EC regulations, from 1995 onwards essential uses will still be allowed providing the user's organization has been issued with a licence from the EC. Lipid analysis qualifies as an essential use. Heat-resistant gloves should be used to handle very hot, and very cold, materials.

Occasionally the risk of inhaling irritant particles or infectious/pathogenic organisms justifies wearing a face mask and of working in a fume hood/cupboard or flow hood/cabinet. Any skin wound should be covered.

The risk of poisoning or infection can be reduced by adopting the following good habits in the laboratory:

(i)   no eating or drinking;
(ii)   no sucking or blowing by mouth of equipment or licking of labels, etc.;
(iii)   no application of cosmetics or ointments;
(iv)   no smoking.

The last point clearly also reduces the risk of fire and is particularly important in lipid work.

Laboratory work should not be undertaken if you are overtired, are under the influence of alcohol or drugs, or in any way feel your performance is impaired. You should not introduce into the laboratory any other person who is not aware of the risks or not able to understand or comply with the safety requirements.

References

1. Kates, M. (1986) *Techniques of Lipidology* (2nd edn). Elsevier, Amsterdam.
2. Meyer, V.R. (1988) *Practical High Performance Liquid Chromatography*. John Wiley, New York.
3. Christie, W.W. (1987) *High Performance Liquid Chromatography and Lipids*. Pergamon Press, Oxford.
4. Blain, R. (1993) in *A Practical Guide to HPLC Detection* (D. Parriott, ed.). Academic Press, London, pp. 39–66.
5. McDonald, P.D. (ed.) (1986) *Waters Sep-Pak® Cartridge Applications Bibliography*. Waters, Division of Millipore, Bedford, MA.
6. Christie, W.W. (1989) *Gas Chromatography and Lipids*. The Oily Press, Ayr (Scotland).
7. Jennings, W. (1987) *Analytical Gas Chromatography*. Academic Press, Orlando, FL.
8. Touchston, J. (1982) *Advances in Thin Layer Chromatography*. John Wiley, New York.
9. Whalen, M.M., Wild, G.C., Spall, W.D. and Sebring, R.J. (1986) *Lipids*, **21**, 267–270.

10. Dyer, J.R. (1965) *Applications of Absorption Spectroscopy of Organic Compounds.* Prentice-Hall, Englewood Cliffs, NJ.
11. Crooks, J.E. (1978) *The Spectrum in Chemistry.* Academic Press, London.
12. Pavia, D.L., Lampman, G.M. and Kriz, G.S. (1979) *Introduction to Spectroscopy.* Saunders College, Philadelphia, PA.
13. Geary, W. (1986) *Radiochemical Methods.* John Wiley, Chichester.
14. Billington, D., Jayson, G.G. and Maltby, P.J. (1992) *Radioisotopes.* BIOS Scientific Publishers, Oxford.
15. Bolton, A.E. and Hunter, W.M. (1986) in *Handbook of Experimental Immunology,* Vol. 1, *Immunochemistry* (M.D. Weir, ed.). Blackwell Science Ltd, Oxford, p. 26.43.
16. Corrie, J.E.T. (1983) *Immunoassays for Clinical Chemistry* (2nd edn). (W.M. Hunter and J.E.T. Corrie, eds). Churchill-Livingstone, Edinburgh.
17. Roitt, I. (1984) *Essential Immunology* (8th edn). Blackwell Scientific Publications, Oxford.
18. Tijssen, P. (1985) *Practice and Theory of Enzyme Immunoassays.* Elsevier, Amsterdam.
19. Hakomori, S. and Kannagi, R. (1986) in *Handbook of Experimental Immunology,* Vol. 1, *Immunochemistry* (M.D. Weir, ed.). Blackwell Science Ltd, Oxford, p. 9.33.
20. Kannagi, R., Levery, S.B. and Hakomori, S. (1983) *Proc. Natl Acad. Sci. USA,* **80,** 2844–2848.
21. Brockerhoff, H. and Yurkowski, M. (1966) *J. Lipid Res.,* **7,** 62–64.
22. Lands, W.E.M. (1965) *Ann. Rev. Biochem.,* **34,** 313–346.
23. Hawkins, M.D. (1980) *Technician Safety and Laboratory Practice.* Cassell, London.
24. Collins, C.H. (1985) *Safety in Biological Laboratories.* John Wiley, Chichester.
25. Lamb, R.A. (1989) *Manual of COSHH Chemical Data for Pathology Laboratories.* Mid Anglia Press, London.
26. Sigma-Aldrich (1994) *Library of Regulatory and Safety Data.* Sigma-Aldrich Co. Ltd, St Louis, MI.
27. Sigma-Aldrich (1994) *Material Safety Data Sheets.* Sigma-Aldrich Co. Ltd, St Louis, MI.

3 Hydrocarbons

3.1 Biological significance

Hydrocarbons are particularly common in higher plants and in insects where they often protect from loss of water (see e.g. ref. 1). They are also frequently found in appreciable quantities in lipid extracts of algae and bacteria. Some polyunsaturated hydrocarbons, the carotenes, play an important role in photosynthesis and are responsible for the orange to red color of several plant products. Having been ingested these pigments may also appear in the fat of animals. Ingested β-carotene may be converted to retinol, or vitamin A, by animal systems, making this compound a valuable dietary component. Unsaturated hydrocarbons such as squalene appear in many eukaryotic cells as intermediates in the biosynthesis of more complex isoprenoid compounds such as cholesterol.

3.2 Structures

Saturated normal hydrocarbons found in nature have the general structure (a) (*Figure 3.1*) in which n = 6–36 or, rarely, greater. Monobranched isomers include the isoforms (b) and the ante-isoforms (c) (*Figure 3.1*).

Those hydrocarbons derived from isoprenoids are multibranched and amongst the most common are farnesane, phytane, pristane and squalane (d, e, f and g, respectively, *Figure 3.1*). They occur in some marine organisms but are found more frequently in geological sediments and in petroleum.

The most common forms of polyunsaturated hydrocarbons containing unconjugated double bonds are squalene (structure a, *Figure 3.2*) and

(a) $CH_3(CH_2)_n CH_3$

(b) $CH_3-(CH_3)CH(CH_2)_{n-2}CH_3$

(c) $CH_3-CH_2(CH_3)CH(CH_2)_{n-3}CH_3$

(d) $H[CH_2-(CH_3)CH-CH_2-CH_2]_3 H$

(e) $H[CH_2-(CH_3)CH-CH_2-CH_2]_4 H$

(f) $H[CH_2-(CH_3)CH-CH_2-CH_2]_3 CH_2-(CH_3)-CH-CH_3$

(g) $H[CH_2-(CH_3)CH-CH_2-CH_2]_3-[CH_2-CH_2-(CH_3)CH-CH_2]_3 H$

FIGURE 3.1: *Structures of selected naturally occurring saturated hydrocarbons. (a) General structure of saturated normal hydrocarbons; (b) isoform monobranched version of a; (c) ante-isoform monobranched version of a; (d) farnesane; (e) phytane; (f) pristane; (g) squalane.*

the isomers of phytadiene, namely 1, 3-, 2, 4- and neo-phytadiene (structures b, c and d, respectively, *Figure 3.2*). The internal double bonds are substituted in a *trans* configuration (methyl to hydrogen group) in phytoene (structure e, *Figure 3.2*). Carotenoids and some of their precursors constitute a group of polyunsaturated hydrocarbons containing series of conjugated *trans* double bonds which in some molecules are sufficiently long to produce intensely red (as in lycopene), orange (as in β-carotene) or yellow (as in α-carotene) pigments. The structures of these three common carotenoids are given in *Figure 3.3* (structures a, b and c, respectively). They are formed naturally from phytoene (structure e, *Figure 3.2*) and other intermediates. An overview of the complex area of carotenoid structures and nomenclature was published recently [2].

3.3 Detection

Carotenoid hydrocarbons are best detected by observation of their orange-red color. Since the presence of one particular carotenoid may be masked by or its detection confused by other pigments, it is probably best to separate a mixture of pigments by extraction and TLC (see Section 3.4). Correspondence of R_f (in several systems) and

(a) $H[CH_3-(CH_3)C = CH-CH_2]_3-[CH_2-CH = (CH_3)C-CH_2]_3 H$

(b) $H[CH_2-(CH_3)CH-CH_2-CH_2]_3-CH = (CH_3)C-CH = CH_2$

(c) $H[CH_2-(CH_3)CH-CH_2 CH_2]_2-CH_2-(CH_3)CH-CH_2-CH = CH-(CH_3)C = CH-CH_3$

(d) $H[CH_2-(CH_3)CH-CH_2-CH_2]_3-CH_2-(CH_2)C-CH = CH_2$

(e) $H[CH_2-(CH_3)C = CH-CH_2]_4-[CH_2-CH = (CH_3)C-CH_2]_4 H$

FIGURE 3.2: *Structures of selected naturally occurring unsaturated hydrocarbons lacking a long system of conjugated double bonds. (a) Squalene; (b) 1,3-phytadiene; (c) 2,4-phytadiene; (d) neo-phytadiene; (e) phytoene.*

(a) $CH_3-(CH_2)C = CH-CH_2-CH_2-(CH_3)C = CH-CH = [CH-(CH_3)C = CH-CH]_2 =$
       $[CH-CH = (CH_3)C-CH]_2 = CH-CH =(CH_3)C-CH_2-CH_2-CH = (CH_3)C-CH_3$

(b)   ![structure] $C = [CH-(CH_3)C = CH-CH]_2 = [CH-CH = (CH_3)C-CH_2]_2 = C$

(c)   as for (b) C

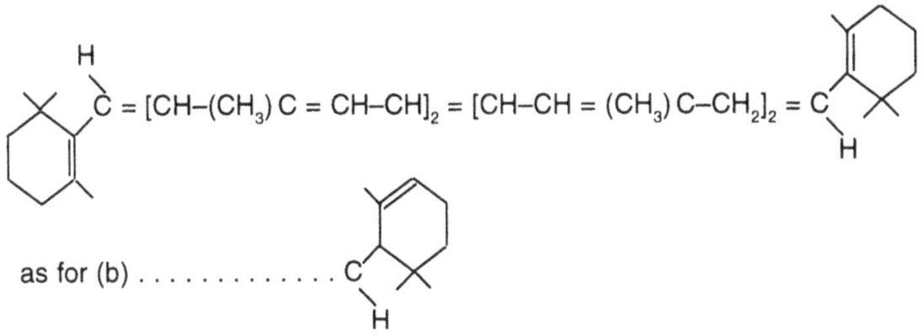

FIGURE 3.3:   *Structures of selected naturally occurring unsaturated hydrocarbons (carotenoids) possessing a long system of conjugated double bonds. (a) Lycopene [ψ,ψ-carotene]; (b) β-carotene [β,β-carotene]; (c) α-carotene [β,ε-carotene].*

color with standards often suffices to detect the presence of a known carotenoid. Apart from the carotenoids, hydrocarbons in general lack chromophoric groups. In this case after TLC the chromatogram may be sprayed with water, allowing the lipid to be seen as a relatively nonwettable area. For greater sensitivity the chromatogram should be sprayed with an aqueous solution of rhodamine G or dichloro-fluorescein or be exposed to iodine vapor. These nondestructive methods are usually sufficiently sensitive for hydrocarbons present in moderate concentrations. For those present in trace quantities (less than 1% total hydrocarbons) the extra sensitivity of the destructive procedure of charring the organic material by heating after spraying with sulfuric acid, aqueous potassium chromate or aqueous ammonium sulfate may well be needed.

3.4 Isolation and purification

Hydrocarbons occur in total lipid extracts and in unsaponifiable lipids (see Section 2.1) derived from most natural sources, although the distribution of particular hydrocarbons may be quite selective (see Section 3.1). Unsaponifiable lipids are easier to analyze and, if protection of unsaturated compounds from oxidation (e.g. by alkaline pyrogallol) and light is provided, recoveries from saponification are good. Analysis of hydrocarbons from total lipid extracts is helped by preliminary precipitation of polar lipids with acetone and recovery of the solute. Again protection of unsaturated compounds from light and from oxidation, preferably by performing as much of the operation in an atmosphere of nitrogen or in the presence of an anti-oxidant, is essential.

Initial separation of hydrocarbons from most other lipids can be conveniently achieved by column adsorption chromatography (LC) on silicic acid using mixtures of carbon tetrachloride/isooctane for elution of the hydrocarbon fraction. Small samples may be easily handled on Sep-Paks®. Eluents are then best tested by TLC.

A popular TLC system involves silica plates and heptane/benzene mixtures as developing solvents. Using a 9:1 (v/v) mixture of heptane/benzene and silica gel H, saturated and monounsaturated hydrocarbons will usually have an R_f above 0.8, whereas isoprenoids and carotenoids are less mobile with R_f values of approximately 0.4 and 0.25 respectively. Silver nitrate-impregnated plates have been used to separate alkenes from alkanes.

Fine separations of hydrocarbon mixtures have been achieved by reversed phase partition HPLC on a C_{18} column fitted with an IR or photometric detector (for carotenoids). *Figure 3.4* illustrates the power of this technique to resolve very closely related lipids. HPLC is the preferred method for separation of carotenoids. Identification of carotenoids eluting from an HPLC column can be assisted by determination of the UV or visible light absorption spectrum. This can be achieved on collected fractions by standard off-line spectrophotometry or by on-line use of a photodiode array detector which allows determination of the absorption spectrum of a series of chromatographic peaks each within 1 sec [4].

Alternatively, GLC of saturated and simple unsaturated hydrocarbons on nonpolar columns such as SE-30 and OV-1 has achieved good separations of homologs and their isomers up to chain lengths of C_{32}. Identification of unsaturated components may be assisted by performing GLC before and after hydrogenation of the sample. A combination of classical chemical oxidation of double bonds using periodate, permanganate, ozonolysis or osmium tetroxide and GLC of the products has proved a useful approach to locating the position of the double bonds in a molecule. However, GC–MS and HPLC–MS are probably the most reliable systems for definitive identification of microgram quantities of a hydrocarbon. If milligram quantities are available, IR and NMR data may confirm an identification.

Capillary GLC columns coated with dimethyl polysiloxane have been used successfully in hydrocarbon analysis. On this nonpolar stationary phase smaller homologs are more mobile than larger ones. Branching also causes delay in elution. The more polar stationary

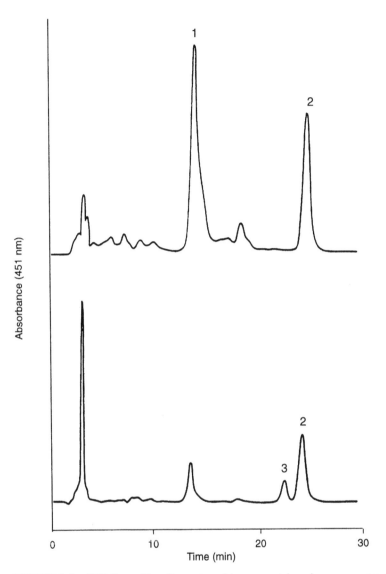

FIGURE 3.4: HPLC profile of carotenoids present in a hexane extract of tomato (**top**) and carrot (**bottom**). Peaks: 1, lycopene; 2, β-carotene; 3, α-carotene. Column: 5 μm ODS C_{18}; mobile phase: acetonitrile–methanol–ethyl acetate (61:14:25 by vol.). A 40 g sample of tissue, pre-extracted with methanol, was extracted with hexane and 4% of the extract (20 μl) was injected directly. Redrawn from ref. 3 with permission from Academic Press.

phases (e.g. carbowax 20M) interact with double bonds, resulting in significant delays in elution times, in proportion to the number of double bonds present. Separation on the basis of number and position of double bonds is usually good. Cyanopropyl polysiloxane columns readily separate geometric isomers, with the E-isomer eluting before the Z-isomer.

3.5 Quantitation

Micro-determination of the amounts of hydrocarbons in natural sources is probably simplest for those that absorb light in the visible or UV range, i.e. primarily carotenoids and their precursors. The complexity of a mixture of carotenoids can be a problem especially for minor components. However, for major components determination of an absorption spectrum or of absorption at selected wavelengths (usually at λ_{max} of the components) on an extract may suffice, especially if TLC indicates the presence of primarily one component. For most mixtures it is best usually to carry out reversed phase HPLC. In many instances most components show base-line separation and recovery of each peak and off-line spectrophotometry is satisfactory. Alternatively, monitoring elution from the column at a selected wavelength and comparison with a calibration curve is sufficient, but this approach relies on retention time and absorption of light at the chosen wavelength for identification, which for a complex mixture may not be sufficient. The use of a sophisticated on-line photodiode array detector to determine the absorption spectrum of each peak provides a powerful micro-assay system [4].

Sensitive assays for saturated and simple unsaturated hydrocarbons are best based on GLC in which micro-detection and quantitation present no difficulties. Less sensitive determination may be based on the weight of a solvent-free chromatographic fraction, especially if TLC confirms the presence of only one component. Densitometry of colored or stained (charred) TLC plates provides a fairly sensitive, perhaps semi-quantitative assay when calibrated against standards run alongside. For carotenoids this method can be unreliable because of fading of the color when plates are exposed to light.

References

1. Kollatukudy, P.E. (1976) *Chemistry and Biochemistry of Natural Waxes.* Elsevier, Amsterdam.
2. Pfander, H. (1992) *Meth. Enzymol.*, **213**, 3–12.
3. Schmitz, H.H., v. Breenen, R.B. and Schwartz, S.J. (1992) *Meth. Enzymol.*, **213**, 322–336.
4. De Leenkeer, A.P. and Nelis, H.J. (1992) *Meth. Enzymol.*, **213**, 251–265.

4 Alcohols, Phenols, Aldehydes, Ketones and Quinones

Since these groups of lipids are closely related biosynthetically and in terms of analytical methodology, it is convenient to deal with them in the same chapter.

4.1 Alcohols and phenols

4.1.1 General and biological significance

Unesterified alcohols that are components of lipid extracts inevitably contain carbon chains that are sufficiently long to be quite hydrophobic. In plants aliphatic alcohols occur as components of cuticular waxes and polyisoprenoid alcohols are found in osmiophilic globules in chloroplasts (and other plastids) and in some cases in woody tissue. Free aliphatic alcohols are not common components of animal systems, although some unsaturated forms are found in insects and have important roles as pheromones. Pristanol and phytanol are found in geological sediments. Polyisoprenoid alcohols in the form of dolichols are present in most tissues of most eukaryotic organisms. In bacteria free aliphatic and polyisoprenoid alcohols occur. When converted to their mono- and diphosphates, fully unsaturated polyisoprenoid alcohols act as intermediate carriers of sugars in bacterial wall polysaccharide synthesis (whilst dolichols function in eukaryotic glycoprotein N-glycan biosynthesis).

Cholesterol is found in all animal tissues at a concentration much higher than that of dolichols and is an important component of the plasma membrane. In liver it functions further as a precursor of bile acids and in some endocrine glands as a precursor of steroid hormones. Analogs of cholesterol (e.g. sitosterol, stigmasterol, ergosterol) are found throughout plants and fungi. Vitamin D_2 and D_3, important in calcium homeostasis in animals, are derived from steroids by UV irradiation.

Many of these alcohols are also present as parts of esters with long chain fatty acids or of ethers with the glycerol portion of phospholipids. The latter are especially common in the archaebacteria.

Retinol, or vitamin A, is an important trace component of several animal tissues, having been formed from β-carotene. Esterified to fatty acids it accumulates in the liver, especially of fish, as a storage form. Small amounts of amino alcohols (sphingosines) can often be found, possibly as hydrolysis products of sphingolipids.

The most common lipid-soluble phenols (chromanols) are the tocopherols (forms of vitamin E). They are found in plant and animal systems where they appear to fulfill an anti-oxidant role. Also found in several plant tissues and in algae are the hydroxylated carotenoids, members of the group of xanthophylls (oxygenated carotenoids), some of which have a role in photosynthesis. Some are also responsible for the color of fruits, flowers, bird feathers, insects and marine animals (e.g. crustacea and salmon), as well as bacteria. An anti-oxidant role for these compounds has also been proposed.

4.1.2 Structures

Saturated normal aliphatic primary alcohols have the general structure (a) (*Figure 4.1*), where n is usually in the range of 6–30 carbon atoms [1]. The corresponding secondary alcohols are much rarer. Monobranched versions of the primary alcohols exist mainly as the isoform and ante-isoform (structures b and c, *Figure 4.1*). Most multibranched saturated alcohols are derived from isoprenoid alcohols, the most common having the general structure (d) in *Figure 4.1,* in which n can be 2 (tetrahydrogeraniol), 3 (farnesanol) or 4 (phytanol). Pristanol (structure e, *Figure 4.1*) is a shortened version of phytanol. A series of monounsaturated, unbranched primary alcohols analogous to the more common series of monoenoic fatty acids (see Chapter 5) also occur in most biological systems, usually in trace amounts. Phytanol is also incorporated by an ether linkage into the more complex alcohols of the membrane-bound ether lipids of archaebacteria. For example, the major membrane lipids of some halobacteria and methanobacteria are derivatives of 2, 3 di-*O*-phytanyl-*sn*-glycerol (structure f, *Figure 4.1*). In some methanobacteria and thermoacidophiles, tetraether derivatives are formed which involve two head to head links between putative terminal

(a) $CH_3(CH_2)_nCH_2OH$

(b) $CH_3(CH_3)CH(CH_2)_{n-2}-CH_2OH$

(c) $CH_3CH_2(CH_3)CH(CH_2)_{n-3}-CH_2OH$

(d) $H[CH_2-(CH_3)CH-CH_2-CH_2]_nOH$

(e) $H[CH_2-(CH_3)CH-CH_2-CH_2]_3-CH_2-(CH_3)CH-CH_2OH$

$H[CH_2(CH_3)CH-CH_2-CH_2]_4-O-C^3H_2$
$|$
(f) $H[CH_2(CH_3)CH-CH_2-CH_2]_4-O-C^2H$
$|$
$HO-C^1H_2$

(g) CH_2OH
$|$
$CH-O[CH_2-CH_2CH(CH_3)-CH_2]_4-[CH_2(CH_3)CH-CH_2-CH_2]_4-O-CH_2$
$|$ $|$
$CH_2-O[CH_2-CH_2CH(CH_3)-CH_2]_4-[CH_2(CH_3)CH-CH_2-CH_2]_4-O-CH$
                                                      $|$
                                                      $HO-CH_2$

FIGURE 4.1: *General formulae and specific structures of some of the more common naturally occurring saturated lipid-soluble alcohols: (a) Saturated normal aliphatic primary alcohol; (b) monobranched isoform of a; (c) monobranched ante-isoform of a; (d) saturated derivatives of isoprenoid alcohols; (e) pristanol; (f) 2, 3-di-O-phytanyl-sn-glycerol; (g) tetraether derivative of f.*

methyl groups of four phytanyl residues each linked to glycerol residues through ether links (structure g, *Figure 4.1*). Usually the major proportion of these phytanyl glycerol derivatives exists as part of amphipathic molecules that carry polar groups on the hydroxyls such as sulfate, phosphate and sugars (see also sections 7.5 and 7.6).

Phytol (structure a, *Figure 4.2*) is probably the best known monoenoic multibranched alcohol, being present in most plants primarily as a hydrophobic ester of chlorophyll. The isoprenoid alcohols with the general structure (b) in *Figure 4.2* can be subclassified on the basis of size (i.e. the value of n) and on the stereochemistry of the substitution of the double bond (methyl relative to the hydrogen atom).

To take account of differences in the number of isoprene residues, their *cis* (Z)–*trans* (E) isomerism (structure d, *Figure 4.2*) and degree of unsaturation, an abbreviated nomenclature has been used in which

(a) $H[CH_2-(CH_3)CH-CH_2-CH_2]_3-CH_2-(CH_3)C=CH-CH_2OH$

(b) $H[CH_2-(CH_3)C=CH-CH_2]_nOH$

(c) $H[CH_2-(CH_3)C=CH-CH_2]_n-CH_2(CH_3)(OH)C-CH=CH_2$

(d)

cis (Z)   trans (E)

(e)

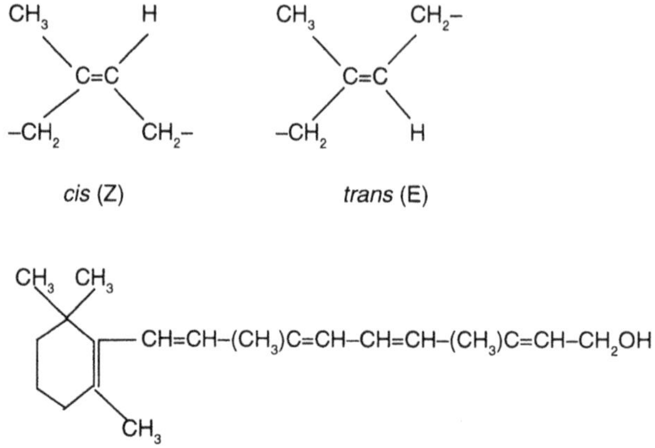

FIGURE 4.2: *Structures of isoprenoid alcohols and derivatives. (a) Phytol; (b) general structure of isoprenoid alcohols; (c) cis (Z) and trans (E) isoprenoid residue; (d) isomeric tertiary alcohol of b; (e) all-trans-retinol (vitamin A₁).*

the letters E (T) and Z (C) represent *trans* and *cis* residues and S a saturated (dihydro) residue, with subscripts to indicate the number of each type of residue. A convention of describing the sequence of residues with the ω-residue to the left and the α-residue to the right is also common. The ω-residue is usually indicated by ω since the substitution of its double bond cannot be E or Z.

In this way the abbreviated representation of the diprenol *trans*-geraniol (*n*=2, structure b, *Figure 4.2*) becomes ωEOH (ωTOH), of the triprenol *trans,trans*-farnesol (*n*=3, structure b, *Figure 4.2*) becomes ωE₂OH (ωT₂OH) and of the tetraprenol all-*trans*-geranylgeraniol (*n*=4, structure b, *Figure 4.2*) becomes ωE₃OH (ωT₃OH). All of these prenols have been described, especially in plant tissues. They are often accompanied by isomeric and saturated versions such as *cis*-geraniol or nerol (ωZOH), ωCOH and dihydrogeraniol or citronellol (ωSOH) in plant essential oils. The isomeric tertiary alcohol versions may also occur in these oils; for example, nerolidol (*n*=2, structure d, *Figure 4.2*) an isomer of farnesol. These are relatively rare and no abbreviation of this form of α-residue has evolved.

Isoprenoid alcohols of five isoprene residues or more are usually described as polyisoprenoid alcohols or polyprenols [2]. Several of the shorter prenols discussed above are present in most cells at low concentrations as hydrolysis products of phosphorylated forms, acting as intermediates *en route* to polyprenols, sterols, polyisoprenoid quinones or, in the plant kingdom, to cyclized forms. Farnesyl and geranylgeranyl residues are also found linked to cysteine residues of prenylated proteins (usually G-proteins) and influence their interaction with membranes (see Chapter 9).

The isoprene residues of some polyprenols are all-*trans* (apart from the ω-residue). The most common of these are solanesol (all-*trans*-polyprenol-9, $\omega E_8 OH$, $\omega T_8 OH$), first identified in tobacco leaves but present in leaves of many plants, and spadicol (all-*trans*-polyprenol-10, $\omega E_9 OH$, $\omega T_9 OH$), first identified in the spadix of *Arum maculation*. It is generally assumed that these alcohols are derived by hydrolysis of the corresponding diphosphates which act as immediate precursors of the polyisoprenoid side chains of plastoquinone-9 and ubiquinone-10 (see Section 4.4).

However, most polyprenols are of mixed stereochemistry. The betulaprenols ($\omega E_2 Z_{3-6} OH$) have been isolated from the woody tissue of trees and the ficaprenols ($\omega E_3 Z_n OH$, $\omega T_3 C_n OH$, n=5–9) have been reported to accumulate in osmiophilic globules in chloroplasts of many green plants. In some green plants the value of n may reach 100, while in some mushrooms up to 300 Z (C)-residues have been reported. Bactoprenols ($\omega E_2 Z_{7-9} OH$, $\omega T_2 C_{7-9} OH$) are slightly longer versions of betulaprenols and are found in bacteria. All eukaryotic organisms contain dolichols ($\omega E_2 Z_n SOH$, $\omega T_2 C_n SOH$, n=10–17), the main features of which are the presence of a saturated, dihydro, α-residue and many Z (C)-residues. From any one natural source, a family of mainly four or five dolichols differing only in the value of n can be isolated. For example, pig liver dolichols are primarily $\omega E_2 Z_{13-17} SOH$ ($\omega T_2 C_{13-17} SOH$) and those from *Saccharomyces cerevisiae* are mainly $\omega E_2 Z_{10-14} SOH$ ($\omega T_2 C_{10-14} SOH$). In some fungi the ω-residue of dolichols may be slightly modified. Both bactoprenols and dolichols have been shown to form mono- and diphosphate derivatives that function as coenzymes in glycosyl transfer (see Chapter 8). The functions, if any, of betulaprenols and ficaprenols are unknown.

An important unsaturated variant of a tetraprenol is retinol (vitamin A) which has a long series of conjugated *trans* double bonds (structure e, *Figure 4.2*), having been formed by oxidative cleavage of β-carotene (see Chapter 3) [3]. The 11-*cis* isomer also occurs naturally in the retina.

There are many hydrophobic cyclic secondary alcohols found primarily in plants and formed mainly by cyclization of mono- and sesquiterpenes (di- and triprenols). This area of knowledge is too specialized for further treatment in this chapter. However, the major components of unsaponifiable lipids of a wide range of eukaryotic organisms are the sterols which are cyclized and substituted derivatives of the triterpene squalene (see Chapter 3) [4]. In animals the main sterol is cholesterol (structure a, *Figure 4.3*) whereas in plants stigmasterol (structure b, *Figure 4.3*) and β-sitosterol (structure c, *Figure 4.3*) predominate. The main member of this group in yeasts and fungi is ergosterol (structure d, *Figure 4.3*). Often a high proportion of sterols is present as fatty acyl esters and sometimes a little appears as glycosides (see Chapter 8).

Important sterol derivatives formed in animals by ultraviolet (UV) irradiation of 7-dehydrocholesterol and of ergosterol are respectively vitamin D_3 (cholecalciferol, structure e, *Figure 4.3*) and vitamin D_2 (ergocalciferol, structure f, *Figure 4.3*) [3]. In animals these are

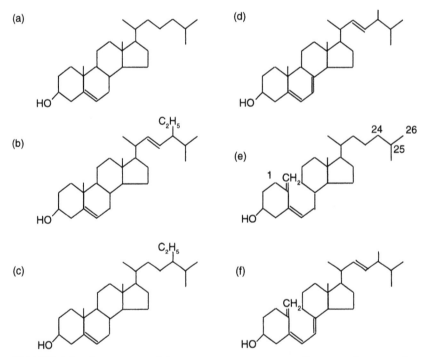

FIGURE 4.3: *Structures of the more common naturally occurring sterols and derivatives. (a) Cholesterol; (b) stigmasterol; (c) β-sitosterol; (d) ergosterol; (e) cholecalciferol or vitamin D_3 (the numbers 1, 24, 25 and 26 indicate the positions at which hydroxylation may take place – see text for details); (f) ergocalciferol or vitamin D_2.*

converted to trace amounts of the 1,25-dihydroxy derivative, which is the active agent in calcium homeostasis, and to 24,25- and 25,26- dihydroxy and 1,25,26- trihydroxy-derivatives which are inactive metabolites.

Steroid hormones rarely occur in lipid extracts of tissues in more than trace quantities and are not dealt with here. The reader is referred to a recent review on this specialized area [5].

Sphingosine is an important noncyclic lipid-soluble compound with primary and secondary hydroxyl groups as well as an amino group as key features (structure a, *Figure 4.4*). It may be accompanied in animal tissues by the saturated derivative dihydrosphingosine (structure b, *Figure 4.4*). Plants contain small quantities of phytosphingosine (structure c, *Figure 4.4*) along with unsaturated versions.

Sphingosines are present naturally mainly as part of more complex lipids (the sphingolipids) in which the amino group is esterified with a fatty acid to form a ceramide (structure d, *Figure 4.4*). Substitution also of the primary alcohol with phosphoryl choline gives rise to sphingomyelin (see Chapter 7 and structure e, *Figure 4.4*) or with one or more monosaccharides provides the glycosphingolipids (see Chapter 8 and structure f, *Figure 4.4*). The free sphingosines in tissues probably result mainly from hydrolysis of the more complex lipids.

A small but important group of fat-soluble phenolic compounds (chromanols) occur as the tocopherols. The general structure of tocopherols is shown in *Figure 4.5* (structure a) [3]. The four methylated derivatives of tocol are α-tocopherol (5,7,8-trimethyltocol, also known as vitamin E and found mainly in animal and plant tissues and seed oils), β-tocopherol (5,8-dimethyltocol), γ-tocopherol (7,8-dimethyltocol) and δ-tocopherol (8-methyltocol). The β-, γ- and δ-tocopherols are found mainly in plant tissues and seed oils. These methylated tocols are often accompanied in plant tissues by the corresponding methylated tocotrienols (α-, β-, γ- and δ-, structure b, *Figure 4.5*), in which the side chain of three saturated residues is replaced by three unsaturated isoprene residues.

Also present in many plant tissues, algae and marine crustaceae are the hydroxylated carotenoids. The structures of six commonly occurring members of this group are given in *Figure 4.6*. Three of these, lutein, neoxanthin and violaxanthin are important components of green leaves and probably have a role in photosynthesis. Fucoxanthin, which occurs mainly in marine algae, is probably the most abundant natural carotenoid. Astaxanthin occurs in crustaceae,

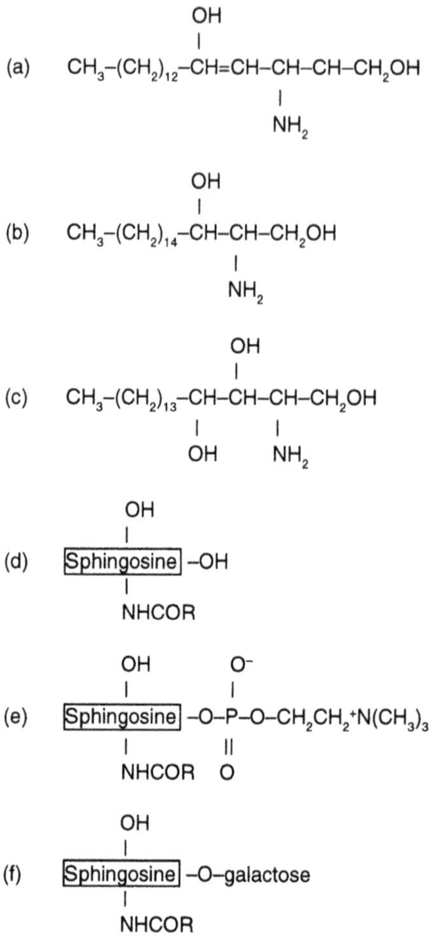

OH
|
(a)   CH$_3$–(CH$_2$)$_{12}$–CH=CH–CH–CH–CH$_2$OH
|
NH$_2$

OH
|
(b)   CH$_3$–(CH$_2$)$_{14}$–CH–CH–CH$_2$OH
|
NH$_2$

OH
|
(c)   CH$_3$–(CH$_2$)$_{13}$–CH–CH–CH–CH$_2$OH
| |
OH   NH$_2$

OH
|
(d)   [Sphingosine] –OH
|
NHCOR

OH   O$^-$
| |
(e)   [Sphingosine] –O–P–O–CH$_2$CH$_2$$^+$N(CH$_3$)$_3$
| ||
NHCOR   O

OH
|
(f)   [Sphingosine] –O–galactose
|
NHCOR

FIGURE 4.4: *Structure of sphingosine and its common derivatives.*
(a) Sphingosine (trans-4-ene); (b) dihydrosphingosine; (c) phytosphingo-
sine; (d) ceramide (COR = fatty acyl group); (e) a sphingomyelin (ceramide-
phosphorylcholine); (f) a glycosphingolipid (galactosyl ceramide).

where it is often present complexed with protein, causing a color
change from red to blue.

4.1.3 Detection

Of the compounds discussed in this section, only the hydroxylated
carotenoids are colored, although most absorb UV light. Nondestructive detection of alcohols and phenols in moderate amounts using
TLC or paper chromatograms can be achieved by spraying with water

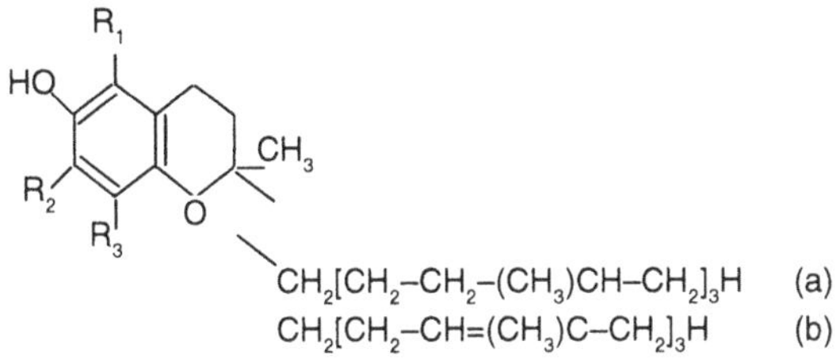

$$CH_2[CH_2-CH_2-(CH_3)CH-CH_2]_3H \quad \text{(a)}$$
$$CH_2[CH_2-CH=(CH_3)C-CH_2]_3H \quad \text{(b)}$$

	R$_1$	R$_2$	R$_3$	
5,7,8-trimethyltocol	Me	Me	Me	α
5,8-dimethyltocol	Me	H	Me	β
7,8-dimethyltocol	H	Me	Me	γ
8-methyltocol	H	H	Me	δ

FIGURE 4.5: *Structures of naturally occurring tocopherols and tocotrienols. (a)-Series = tocols (tocopherols); (b)-series = tocotrienols. (aα) α tocopherol; (aβ) β-tocopherol; (aγ) -tocopherol; (aδ) δ-tocopherol; (bα) α-tocotrienol; (bβ) β-tocotrienol; (bγ) γ-tocotrienol; (bδ) δ-tocotrienol.*

or an aqueous solution of rhodamine G or dichlorofluorescein or by exposure to iodine vapor and viewing in visible light. Viewing under UV light is particularly useful for sensitive detection of UV-absorbing compounds such as carotenoids, retinol and the tocopherols and tocotrienols. UV light of low wavelength will assist in detecting compounds with two conjugated double bonds such as the forms of vitamin D. Retinol fluoresces a bright golden color at 256 nm. The forms of vitamin D fluoresce at 365 nm and the tocopherols at 340 nm after excitation at 295 nm.

Several sensitive destructive methods have been developed for detection with TLC. All of these compounds can be detected nonselectively by charring by heating after spraying with sulphuric acid, aqueous potassium chromate or aqueous ammonium sulfate. Spraying with a mixture of anisaldehyde–sulfuric acid–ethanol–water (5:5:81:9 by vol.) followed by heating produces characteristic colors with isoprenoid compounds. For example, polyprenols stain blue-green, tocopherols stain yellow and sterols and phytol stain blue. The color may vary sligtly with the amount of isoprenoid. An ethanolic solution of phosphomolybdic acid (1:4 w/v) produces a deep blue color when heated with most lipids. Reducing agents such as the tocopherols give a blue color without heating. Tocopherols can also be detected by their reduction of a freshly prepared mixture of ferric chloride (0.2% by vol.) and 2:2'-dipyridyl (0.5% by vol.) in ethanol to give bright red spots at room temperature. Replacement of 2:2'-dipyridyl by 4:7-diphenyl-1:1 phenanthroline increases sensitivity to

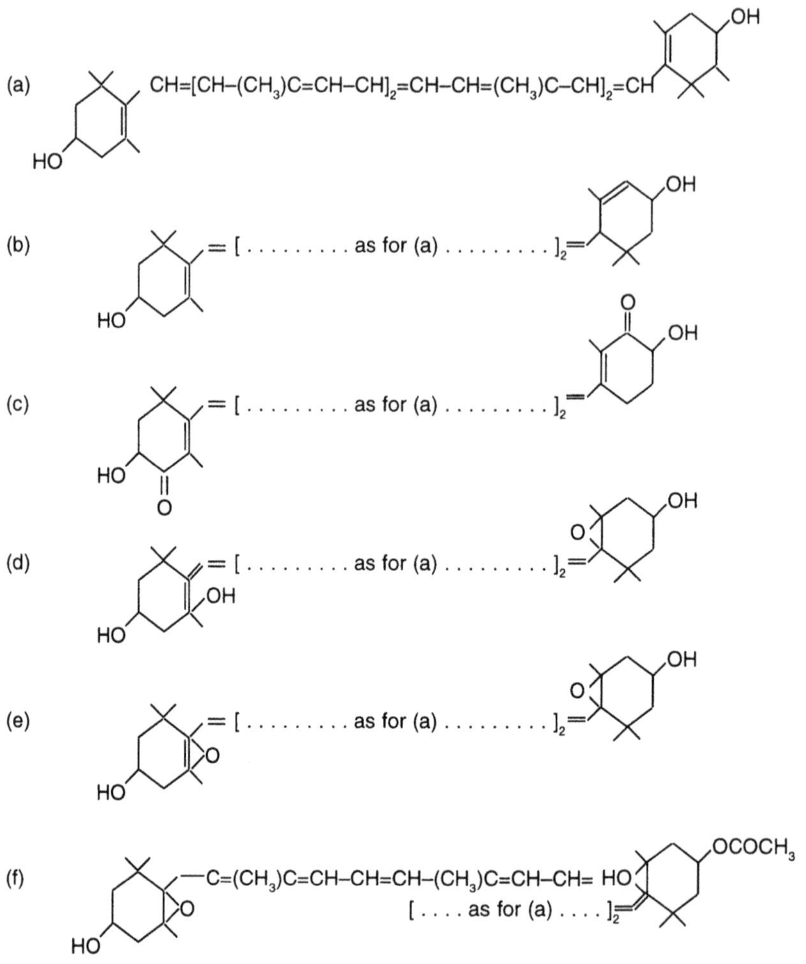

FIGURE 4.6: *Structures of some of the more common hydroxylated carotenoids. (a) Zeaxanthin [(3R, 3'R)-β,β-carotene-3, 3'-diol]; (b) lutein [(3R,3R')-β-carotene-3, 3'-diol]; (c) astaxanthin [(3S,3'S)-3, 3'dihydroxy-β, β-carotene-4, 4'-dione]; (d) neoxanthin [(3S, 5R, 6R, 3'S, 5'R, 6'S)-5', 6'-epoxy-6, 7-didehydro-5, 6, 5'6'-tetrahydro-β, β-carotene-3, 5, 3' triol]; (e) violaxanthin [(3S, 5R, 6S, 3'S, 5'R, 6'S)-5, 6, 5', 6',-diepoxy-5, 6, 5', 6'-tetrahydro-β, β-carotene, 3-3'diol]; (f) fucoxanthin [(3S, 5R, 6S, 3'S, 5'R, 6'R)-5, 6-epoxy-3, 3', 5'-trihydroxy-6'7'-didehydro-5, 6, 7, 8, 5', 6'-hexahydro-β, β-caroten-8-one 3'-acetate].*

less than 1 µg tocopherol. The Liebermann–Burchard reaction forms the basis of a reasonably sensitive detection of cholesterol and its esters. The reagent consists of concentrated sulfuric acid–acetic anhydride (1:4 v/v). After a few minutes a green color develops which slowly changes to blue. If sprayed with a saturated solution of anti-mony trichloride in chloroform containing a few drops of acetic anhydride and heated TLC plates will reveal the presence of retinol and its esters as a blue spot (the Carr–Price reaction), whilst the forms of vitamin D give brown spots and cholesterol a violet color. Particular care should be taken with all of these spray reagents, some of which are very toxic and/or very corrosive.

4.1.4 Isolation and purification

Unesterified and esterified alcohols will be present in total lipid extracts. If it is important to separate the unesterified form, an initial preparative chromatography on silica or alumina columns will be necessary. For example, esters of polyisoprenoid alcohols, cholesterol and retinol will be eluted from silica and from alumina (Brockmann Grade III) by diethylether–hexane (1:99 v/v), whereas elution of the unesterified compounds requires a more polar mixture (between 1:10 and 1:5 v/v). Saponification of the eluted fractions will then release the required alcohols from esters and render the alcohols free of some contaminating esters. Protection from light and prevention from oxidation by the presence of pyrogallol is essential during this step. Recovery of total alcohol (esterified plus unesterified) is probably best achieved by starting with the unsaponifiable lipid obtained by sap-onification of the tissue. However, in the case of plant tissues that contain a high content of insoluble polysaccharides such as cellulose, it may be preferable to first extract the total lipid and then prepare the unsaponifiable lipid from this extract. In the absence of a saponification step, precipitation of polar lipids by acetone may be helpful as a preliminary step to chromatography. Further purification of the alcohol fractions will probably require chromatography on silica or alumina columns eluting with a shallow gradient of increasing concentration of diethyl ether in hexane. Chromatography of 10 mg of lipid on 1 g of adsorbent with a stepwise gradient of 10 ml of each eluent increasing by 2% diethyl ether in hexane at each step results in the lipids normally elutin in the following order: paraffinic alcohols and dolichols, phytol, polyprenols, monohydroxycarotenoids, toco-pherols, phytanylglycerols, retinol, cholesterol, dihydroxy carotenoids and sphingosine. The exact order of elution depends upon the actual chain length and degree of saturation of the compounds in each of these families.

Uniquely, unesterified cholesterol may be precipitated quantitatively from an ethanolic solution of total lipid extract by digitonin (0.5% w/v). A solution of the complex (e.g. in acetic anhydride) can then be assessed directly for cholesterol. Alternatively, if the object of the

exercise is to prepare dolichols and/or polyprenols from unsaponifiable lipids a convenient early step is to use C_{18} Sep-Pak® cartridges,
eluting the major contaminant, cholesterol, with methanol and the
dolichols/polyprenols with ethanol.

Often reversed phase partition chromatography of the eluted
fractions will achieve complete separation and purification of
individual alcohols. This is particularly so if reversed phase HPLC is
used. An example is given in *Figure 4.7* in which cholesterol and
dolichols of a rat liver extract are completely separated from each
other and from ubiquinones-9 and -10 (see Section 4.4) and esters of
the extract as well as from internal standards of ergosterol and
ubiquinone-6. One disadvantage of relying primarily on reversed
phase HPLC for purification of dolichols is the overlap between
dolichols and fully unsaturated polyprenols of similar chain length. If
this is likely to be a problem it is best to separate dolichols from the
corresponding unsaturated polyprenols by adsorption chromatography (dolichols elute first) before reversed phase HPLC.

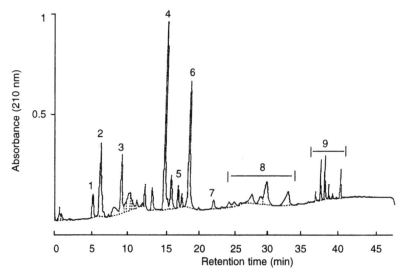

FIGURE 4.7: *Reversed phase HPLC of neutral lipids of rat liver using a
3 μm C-18 column and a convex gradient of methanol–2-propanol–n-
hexane (2:1:1 by vol.) running into methanol–water (9:1 by vol.) at 1.5 ml
min⁻¹. Detection was at 210 nm. Peaks: 1, ergosterol; 2, cholesterol; 3,
ubiquinone-6; 4, ubiquinone-9 (red); 5, ubiquinone-10 (red); 6, ubiquinone-
9; 7, ubiquinone-10; 8, cholesteryl esters and triglycerides; 9, dolichols-17,
-18, -19, -20, -21, -22, -23 (in order of increasing retention time).
Ergosterol, ubiquinone-6 and dolichol-23 were added to the extract as
internal standards. Analysis of ubiquinones is discussed in Section 4.4.
Redrawn from ref. 5 with permission from Elsevier Science Ireland Ltd,
Bay 15K, Shannon Industrial Estate, Co. Clare, Ireland.*

Release of sphingosine (and its derivatives) from the N-acyl groups present in sphingolipids requires more severe hydrolytic conditions such as 2 M hydrochloric acid in methanol at 75°C for 5 h. This treatment also removes substituents (sugars, phosphorylcholine, etc.) from the primary hydroxyl group of the sphingosine. The liberated hydrophobic amino alcohol can be further purified and identified by conversion to dinitrophenyl derivatives (using 1-fluoro-2,4-dinitrobenzene) and use of column chromatography on silicic acid followed by TLC. Alternatively GLC or GC–MS of the N-acetyl, O-TMS (O-trimethylsilyl) derivatives on SE-30 has proved successful. Another useful method is the periodate oxidation of the alcohol to the aldehyde and further conversion of this to the dimethyl acetal. This can then be analyzed by GLC on Apiezon L columns.

4.1.5 Quantitation

Those compounds that absorb visible or UV light may be assayed most readily by spectrophotometry. Depending on the quantity of material present and the complexity of the mixture this may be applied directly to the total lipid extract or after chromatographic separation. As indicated in Section 4.1.3, these approaches are relevant in the visible region particularly to the hydroxylated carotenoids. For example in hexane, zeaxanthin shows strong absorption at 483, 451 and 423 nm, lutein at 477, 447 and 420 nm, violaxathin at 472, 443 and 417.5 nm, and fucoxanthin at 469, 451 and 438 nm, all having molar extinction coefficients (E_{mol}) in the range 1 x 10^5–1.5 x 10^5 at the peak between 440 and 455 nm. Tocopherols can be assayed spectrophotometrically in the UV region but with much lower sensitivity (E_{mol} = 3 x 10^3–4 x 10^3 at 290–300 nm). On the other hand, the fluorescence of retinol and of the forms of vitamin D and the tocopherols described in Section 4.1.3 can form the basis of a sensitive assay for these vitamins.

A routine sensitive assay for tocopherols relies on their reduction of ferric chloride in the presence of 2:2'-dipyridyl to give a red color (Emmerie–Engel reaction) with a λ_{max} at 520 nm. Sphingosine and related amino alcohols may be determined colorimetrically by reaction of the amino group with Ninhydrin reagent and measurement of the red color at 575 nm (E_{mol} = 19 x 10^3). Other routine quantitative determinations dependent upon color reactions are the Liebermann–Burchard reaction for cholesterol (λ_{max} = 620 nm) and the Carr–Price reaction for retinol (λ_{max} = 617 nm). Cholesterol can also be often estimated conveniently by an enzymatic method utilizing cholesterol oxidase which generates hydrogen peroxide by reaction 1 of *Figure 4.8*. Various colorimetric assays of the resultant hydrogen peroxide have been developed. One of these utilizes

Reaction 1:

$$Cholesterol + O_2 \longrightarrow \Delta4\text{-cholestenone} + H_2O_2$$

Reaction 2:

$$2H_2O_2 + 4\text{-aminophenazone} + phenol \longrightarrow 4H_2O +$$
4-(*p*-benzoquinone - monoimino) - phenazone

FIGURE 4.8: *Reactions involved in enzymatic assay of cholesterol. Reaction 1 is catalyzed by cholesterol oxidase and reaction 2 by peroxidase. The intensity of color of the final product is measured at 500 nm.*

peroxidase and an excess of the substrates 4-aminophenazone and phenol, which together produce the colored derivative indicated in reaction 2 of *Figure 4.8*. The intensity of color (λ_{max} = 500 nm, $E_{mol} \approx$ 7 x 10^3) is directly proportional to the amount of cholesterol present. The method is specific to unesterified cholesterol. Cholesteryl esters may first be hydrolyzed by treating with cholesterol esterase. Such methods form the basis of many commercially available test kits for the quantitation of free and total cholesterol in plasma.

Probably the most versatile and universal assay procedure for hydrophobic alcohols possessing one or more double bonds is HPLC, particularly reversed phase HPLC. *Figure 4.7* demonstrates the sensitive quantitation possible (as well as identification of components) when a flow-through UV detector set at 210 nm is employed.

The smaller molecular weight, heat stable members of this group may be measured with very high sensitivity using GLC after conversion to volatile derivatives . This has been employed successfully for sterols and isoprenoid alcohols containing fewer than nine isoprene residues. It has also been the method of choice for quantitation of sphingosine derivatives.

4.2 Aldehydes and ketones

4.2.1 General and biological significance

Hydrophobic aldehydes occur widely in nature, although amounts of the free entities in living organisms are usually very small. Somewhat higher concentrations can be found in lipid extracts after acid hydrolysis due to their release from the aldehydogenic lipids, the plasmalogens (see Chapter 7). Possibly the biologically most important aldehydes are of isoprenoid origin. These include the volatile odorous components of essential oils of plants and insect pheromones. They also include retinal, the active form of vitamin A in vision which condenses with the protein opsin to give the visual pigment rhodopsin. An analogous compound, bacteriorhodopsin, functions in the proton pump of Halobacteria.

Small quantities of paraffinic aldehydes and ketones can be found in plant waxes, insect cuticles and in micro-organisms. They are often responsible for the flavors and odors of foods. Some branched and cyclic ketones show pheromone activity. Important carotenoids carrying keto groups are astaxanthin and fucoxanthin (see structures c and f, respectively, *Figure 4.6*). Several steroid hormones have keto groups essential for their biological activity. Although they will generally be lipid soluble, further discussion of their analysis is beyond the scope of this book.

4.2.2 Structures

The general formulae and structures of some lipid-soluble aldehydes and ketones are summarized in *Figure 4.9*. Variations on the general ketone formulae (*Figure 4.9*, f and g) occur in the form of mono-unsaturation in some bacterial products. The presence of branch methyl groups on some insect pheromones and of cyclization in some animal sex pheromones is another modification that has been reported.

4.2.3 Detection

Most of these compounds can be detected using the nondestructive methods summarized in Section 4.1.3 for related alcohols. Detection methods specific to the aldehyde or ketone groups depend on condensation with colored reagents such as *p*-nitrophenylhydrazine or 2,4-dinitrophenylhydrazine.

Plasmalogens, aldehydes or ketones on TLC plates can be revealed as yellow spots on spraying with an acid solution of 2,4-dinitrophenylhydrazine. Subsequent spraying of the plate with a dilute solution of potassium ferricyanide in hydrochloric acid gives an immediate blue color with ketones, whereas a green color is formed slowly with aldehydes. Schiff's reagent *p*-rosaniline (fuchsin) in sodium bisulfite gives a pink-mauve stain with adehydes on TLC plates but no color with ketones, also allowing the two to be distinguished from each other. Detection of plasmalogens by Schiff's

(a) $CH_3(CH_2)_nCHO$

(b) $CH_3(CH_2)_aCH=CH(CH_2)_bCHO$

(c) 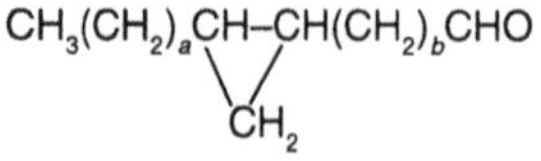 $CH_3(CH_2)_aCH-CH(CH_2)_bCHO$
                      $\diagdown\diagup$
                       CH_2

(d) $H[CH_2(CH_3)C=CHCH_2]_nCH_2(CH_3)C=CHCHO$

(e) 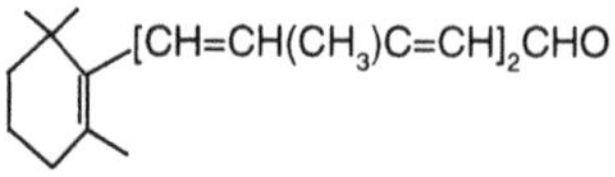 $[CH=CH(CH_3)C=CH]_2CHO$

(f) $[CH_3(CH_2)_n]COCH_3$

(g) $[CH_3(CH_2)_n]_2CO$

FIGURE 4.9:   *General formulae and structures of some naturally occurring lipid-soluble aldehydes and ketones. (a) Saturated normal aliphatic aldehyde, n = 6–21; (b) monoenoic normal aliphatic aldehyde, a = 0–11, b = 0–13, a+b = 0–20; (c) cyclopropane aliphatic aldehyde, a = 5–7, b = 6–9, a+b = 12–14; (d) isoprenoid aldehydes – geranial, n = 1, farnesal, n = 2, geranyl geranial, n = 3; (e) retinal; (f) methyl aliphatic ketones, n = 5–20; (g) symmetrical aliphatic ketones, n = 3–16.*

stain requires the presence of mercuric chloride in the reagent. This slowly hydrolyzes the vinyl–ether linkage so that after approximately 10 min a mauve color develops.

The mobility of these compounds in TLC systems may also be an important aspect of their detection. In this respect most paraffinic ketones, aldehydes and neutral plasmalogens may be distinguished using petroleum ether – ether – acetic acid (approximately 90:10:1) as mobile phase and silica gel H plates. In this system, these compounds chromatograph with an R_f range of approximately 0.5–0.7. Identification of aldehydes and ketones by GLC on a polar column such as butanediol succinate can be successful, although their retention times may be close. However, GC–MS largely removes this problem due to characteristic cracking patterns.

4.2.4 Isolation and purification

In general the approaches discussed in Section 4.1.4 for alcohols are also successful with aldehydes and ketones, bearing in mind that the corresponding alcohols will be rather more polar and therefore less mobile on adsorption chromatography. On reversed phase chromatography the greater polarity of alcohols has much less effect on mobility.

4.2.5 Quantitation

Again the general methods discussed in Section 4.1.5 for alcohols may be applied to this group of compounds also. The keto-carotenoids (e.g. fucoxanthin) exhibit characteristic and powerful absorption of visible light and retinal shows strong absorption of UV light (λ_{max} = 381 nm, E_{mol} = 43.5 x 10^3 in ethanol) which can form the basis of sensitive determination. The Carr–Price reaction described in Section 4.1.5 for the assay of retinol at 617 nm may also be used for assay of retinal at 574 nm.

The use of chromophoric condensation reagents discussed in Section 4.2.3 forms the basis of a colorimetric assay of aldehydes and ketones. For example, products of condensations with p-nitrophenylhydrazine have a λ_{max} of 395 nm and E_{mol} of approximately 24 x 10^3.

The estimation of aldehydes and ketones by GLC is straightforward since they are relatively volatile without having to form a derivative. However, it is usually recommended that, because of their relative instability, the aldehydes are converted into more stable, volatile derivatives such as dimethyl acetals (by heating with methanolic hydrochloric acid), alcohol acetates (by reduction with lithium aluminum hydride followed by acetylation) or methyl fatty acids (by oxidation with chromium trioxide followed by methylation). GLC of any of these products allows identification and determination of most common aldehydes and ketones and their derivatives with double bonds, cyclopropane rings and branch methyls.

4.3 Quinones

4.3.1 General and biological significance

All of the major hydrophobic quinones of natural origin contain polyisoprenoid side-chains which anchor them in membranes where the quinone performs its function. This involves the transfer of electrons involving reversible reduction and oxidation, probably through the corresponding quinol (hydroquinone). Thus ubiquinone functions in electron transport in the mitochondria of all eukaryotic organisms and in the membranes of several bacteria. In some bacteria the function of ubiquinone is taken over by menaquinones. In higher plants photosynthetic electron transport in chloroplasts involves plastoquinone. There are also small quantities of a menaquinone known as phylloquinone in this organelle. Plant and bacterial menaquinones also have an important role as vitamins K_1 and K_2; they cannot be synthesized by animals, yet they play an important role in blood clotting. They undergo reduction to the hydroquinol which is an essential co-factor in the carboxylation of glutamate residues to γ-carboxyglutamate in the precursor proteins of several blood clotting factors. This carboxylation is critical to the blood clotting function of these factors.

4.3.2 Structures

Two general groups of hydrophobic quinones exist: the benzoquinones and the naphthoquinones. The former are further subdivided into the ubiquinones and the plastoquinones. Each of these benzoquinones exists as a family of almost identical compounds, differing only in the number of isoprenoid residues in a long hydrophobic all-*trans*-polyisoprenoid side-chain. Their structures are given in *Figure 4.10* (structures a and b). One of the naphthoquinones (phylloquinone) exists as a single molecular entity with a hydrophobic side-chain made up of the phytyl carbon skeleton (structure c, *Figure 4.10*). The others exist as a family of almost identical compounds, again differing only in the number of isoprenoid residues (structure d, *Figure 4.10*).

Also present in some lipid extracts are the tocopherolquinones generated by oxidation of the tocopherols. For example, α-tocopherolquinone (structure e, *Figure 4.10*) is an oxidation product of α-tocopherol. Quinones corresponding to the β-, γ- and δ-tocopherols (i.e. the 2,5 di-, the 2,3 di- and the 2-monomethyl 1,4-benzoquinones) with a long side-chain on position 6 in all cases, have been described. It is not clear what proportion of these quinones is of natural origin and what is due to post-extraction oxidation.

4.3.3 Detection

In solution each of these quinones shows characteristic absorption of UV light, much of which is shifted and decreased in intensity by reduction with sodium borohydride to the quinol (*Figure 4.11, Table 4.1*). Most lipid extracts require prior partial chromatographic purification of quinones from other chromophoric compounds before this detection method plus chromatographic properties can be used for identification. On chromatography on alumina or silica these quinones will usually run in the order (high to low mobility): menaquinones, plastoquinones, ubiquinones and tocopherolquinones between (approximately) the triacylglycerols and the sterols, depending on the solvent system employed. On TLC plates quinones show up as dark spots (bands) in UV light, especially if run on a

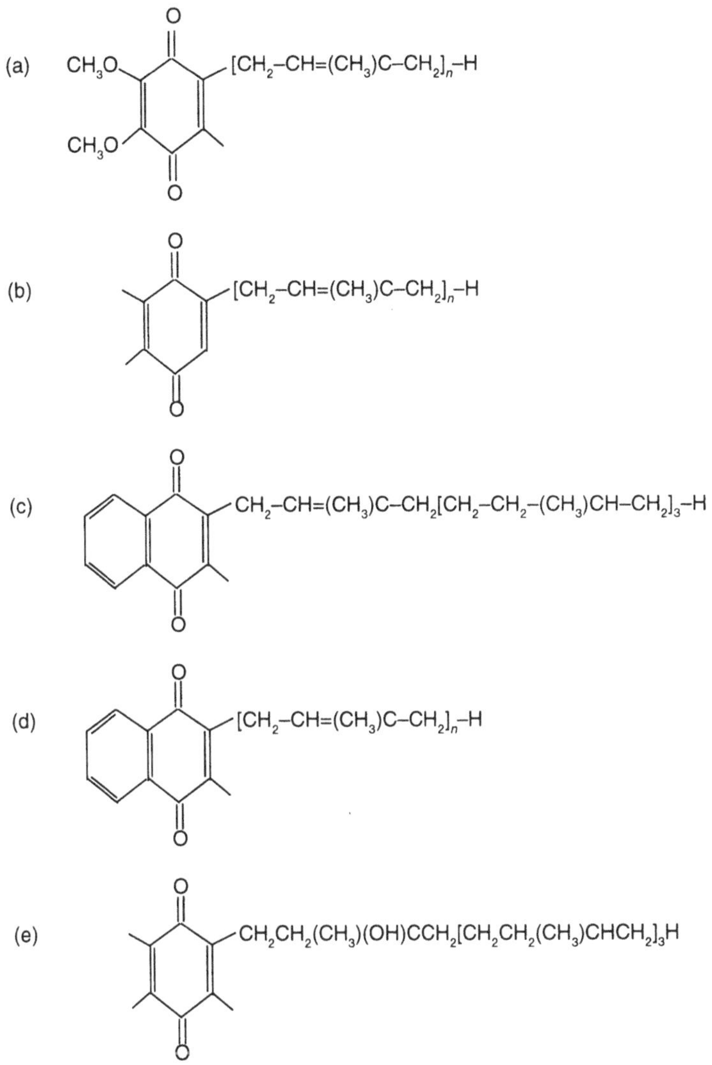

(a) CH_3O — [...] — $[CH_2-CH=(CH_3)C-CH_2]_n-H$

(b)   — $[CH_2-CH=(CH_3)C-CH_2]_n-H$

(c)   — $CH_2-CH=(CH_3)C-CH_2[CH_2-CH_2-(CH_3)CH-CH_2]_3-H$

(d)   — $[CH_2-CH=(CH_3)C-CH_2]_n-H$

(e)   — $CH_2CH_2(CH_3)(OH)CCH_2[CH_2CH_2(CH_3)CHCH_2]_3H$

FIGURE 4.10: *Structures of some naturally occurring lipid-soluble quinones. (a) Ubiquinones (n = 6–10, e.g. in ubiquinone-8 or Q-8, n = 8) or coenzymes Q, 2,3-dimethoxy-5-methyl-6-polyprenyl 1,4-benzoquinone; (b) plastoquinones (n = 6–9, e.g. in plastoquinone-9 or PQ-9, n = 9), 2,3-dimethyl-6-polyprenyl-1, 4-benzoquinone; (c) phylloquinone (K) or vitamin K₁, 2-methyl-3-phytyl-1, 4-napthoquinone; (d) menaquinones (n = 6–10, e.g. in menaquinone-10 or MK-10, n = 10) or vitamin K₂ , 2-methyl-3-polyprenyl-1,4-naphthoquinone; (e) α-tocopherolquinone, 2,3,5-trimethyl-6-(2-hydro, 3-hydroxy) phytyl 1,4-benzoquinone.*

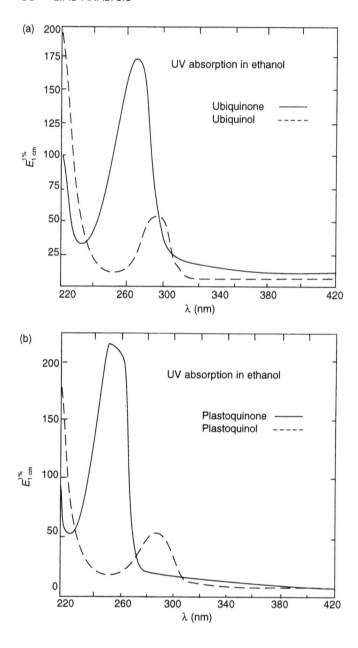

FIGURE 4.11:   *UV absorption spectra of ethanolic solutions of ubiquinone*
(a) and plastoquinone (b) before (——) and after (- - -) reduction with
sodium borohydride to the corresponding quinol. Redrawn from ref. 6
with permission from Academic Press.

TABLE 4.1: *UV absorption spectra of naturally occurring hydrophobic quinones in ethanol*

	Quinone		Quinol	
	λ_{max} (nm)	E_{mol} (approx.)	λ_{max} (nm)	E_{mol} (approx.)
Ubiquinones	275	15×10^3	290	4.5×10^3
Plastoquinones	254	15×10^3	290	3.5×10^3
	261	14×10^3		
Menaquinones	242	18×10^3	244	3.3×10^3
	248	19×10^3	325	3.5×10^3
	260	17×10^3		
	268	17×10^3		
α-Tocopherol	262	16×10^3	286	
Quinone	269	16×10^3		

fluorescent absorbent or sprayed with fluorescein. The different isoprenologs run together on adsorption chromatography but on reversed phase partition they can be readily separated from each other. Reversed phase HPLC has been used very successfully in detecting quinones, quinols and their individual isoprenologs. The application of this approach to ubiquinones and ubiquinols of animal tissues is illustrated in *Figure 4.7*.

Care to protect quinones from oxidative, alkaline and photo-degradation needs to be taken during analysis. If tissues have to be saponified, oxygen should be eliminated from the system and saponification should be performed in the dark. This is best achieved by adding pyrogallol to the saponification mixture and by carrying out the procedure under a blanket of nitrogen. Lipid extraction is also best carried out in an atmosphere of nitrogen, with samples kept in the dark and in the presence of anti-oxidants. This is particularly so if quinols are not to be oxidized to quinones. TLC of quinols is also possible under an atmosphere of nitrogen. The presence on TLC plates of quinols (which are basically reducing agents) is readily detected by methods suitable for tocopherols, such as using the Emmerie–Enyel reagent (ferric chloride–2:2 dipyridyl; see Section 4.1.5).

4.3.4 Isolation and purification

Chromatography on columns of alumina has been used for many years in the isolation and purification of hydrophobic quinones.

Usually the tissue is first saponified, taking the precautions mentioned in Section 4.3.3, and the unsaponifiable lipid is chromatographed using increasing proportions of diethyl ether in petroleum, as described in Section 4.1.4. The relative chromatographic mobility of these quinones has been described in Section 4.3.3.

Reversed phase HPLC has been used successfully to isolate quinones and quinols from lipid extracts of animal tissues (see e.g. *Figure 4.7*).

4.3.5 Quantitation

The characteristic UV absorption spectra of these quinones and quinols has already been discussed in Section 4.3.3 (see also *Figure 4.11* and *Table 4.1*). Of further particular use is the change in absorption upon reduction of quinone to quinol (or oxidation in the reverse direction). For example, the change (ΔE_{mol}) in absorption at 275 nm in ethanol for ubiquinones upon reduction with a small crystal of sodium borohydride is approximately 12.5×10^3 and, providing no other substance is present which absorbs light at 275 nm and is affected by borohydride, this provides a specific and quite sensitive assay for total ubiquinones in a solution. To assay for the isoprenologs of ubiquinone they must first be separated by reversed phase partition chromatography. This can be by reversed phase HPLC or reversed phase TLC followed by recovery of the separated quinones and spectrophotometric assay as above. Alternatively, direct assay by reversed phase HPLC at 210 nm as illustrated in *Figure 4.7* is probably easier.

References

1. Kolattakudy, P.E. (1976) *Chemistry and Biochemistry of Natural Waxes*. Elsevier, Amsterdam.
2. Hemming, F.W. (1974) in *Biochemistry of Lipids* (T.W. Goodwin, ed.). Butterworths, London, pp. 39–97.
3. De Luca, H.F. (1978) in *Handbook of Lipid Research*, Vol 2. Plenum Press, New York.
4. Law, J.H. and Rilling, H.C. (eds) (1985) *Meth. Enzymol.*, **100.**
5. Aberg, F., Zhang, Y., Appelkvist, E.-L. and Dallner, G. (1994) *Chem. Biol. Interact.*, **91,** 1–14.
6. Morton, R.A. (1965) *The Biochemistry of Quinones*. Academic Press, London.

5 Fatty Acids and Prostaglandins

5.1 Fatty acids

5.1.1 Types of fatty acids found naturally

Most of the long chain fatty acids found in living tissues have even numbers of carbon atoms. They can be divided into three groups according to their degree of unsaturation: saturated, monounsaturated and polyunsaturated. A convenient shorthand nomenclature gives the total number of carbons in the chain, followed by the number of double bonds. For instance, linoleic acid (*Figure 5.1*) would be 18:2. The systematic name is *cis*-9,12 octadecadienoic acid, giving the position of the double bonds counting from the carboxyl. The double bonds have the *cis* configuration. In most of the polyunsaturated acids the double bonds are separated by one methylene (CH_2) group. *Tables 5.1* and *5.2* list the more common fatty acids.

Less common fatty acids include those with a ring system (*Figure 5.1*), which can be 3-carbon (cyclopropane or cyclopropene) or 5-carbon (cyclopentene). The cyclopropane fatty acids are found in bacteria, the others in seed oils. Hydroxy acids are also relatively rare, though castor bean oil is rich in 12-hydroxy oleic or ricinoleic acid. A 2-hydroxy derivative of the 24:0 lignoceric acid is found in brain cerebrosides and is known as cerebronic acid.

Branched chain fatty acids are also uncommon, but widely distributed (bacteria, marine oils, animal fats). A methyl group usually forms the branch and is situated near the end of the chain (*Figure 5.1*). Multi-branched acids such as phytanic occur in micro-organisms.

5.1.2 Separation methods for fatty acids

Free fatty acids are found only in small amounts in living tissues. They may be separated from other lipid classes in a chloroform–

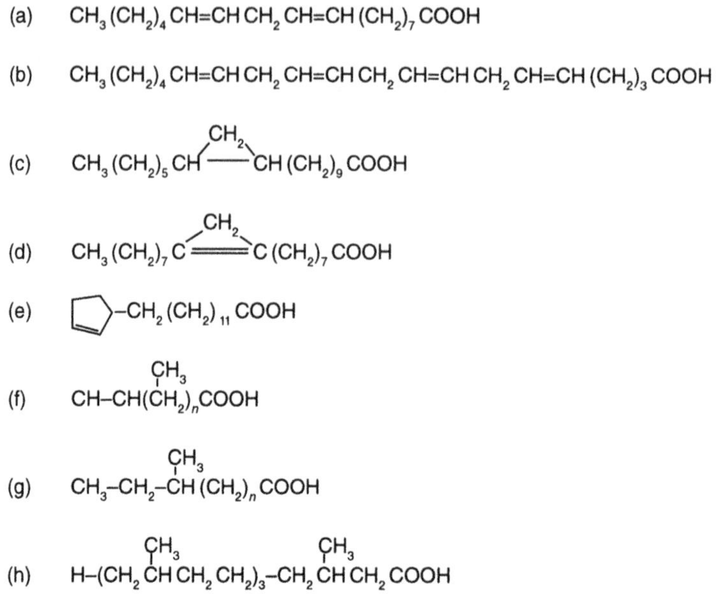

(a) $CH_3(CH_2)_4 CH=CH CH_2 CH=CH(CH_2)_7 COOH$

(b) $CH_3(CH_2)_4 CH=CH CH_2 CH=CH CH_2 CH=CH CH_2 CH=CH(CH_2)_3 COOH$

(c) $CH_3(CH_2)_5 CH\overset{\overset{\displaystyle CH_2}{\diagdown\diagup}}{\text{———}}CH(CH_2)_9 COOH$

(d) $CH_3(CH_2)_7 C\overset{\overset{\displaystyle CH_2}{\diagup\diagdown}}{=\!=\!=}C(CH_2)_7 COOH$

(e) $\langle\!\Box\rangle\!-\!CH_2(CH_2)_{11} COOH$

(f) $CH-CH\overset{\overset{\displaystyle CH_3}{|}}{(CH_2)_n}COOH$

(g) $CH_3-CH_2-\overset{\overset{\displaystyle CH_3}{|}}{CH}(CH_2)_n COOH$

(h) $H-(CH_2 \overset{\overset{\displaystyle CH_3}{|}}{CH} CH_2 CH_2)_3-CH_2 \overset{\overset{\displaystyle CH_3}{|}}{CH} CH_2 COOH$

FIGURE 5.1:   *Structures of fatty acids. (a) Linoleic; (b) arachidonic; (c) lactobacillic; (d) sterculic; (e) chaulmoogric; (f) branched iso series; (g) branched ante-iso series; (h) phytanic.*

methanol extract by adsorption chromatography on a column of silicic acid (silica gel). Elution is by increasingly polar solvents. Kates [1] illustrates the method with hexane–diethyl ether mixtures containing up to 15%(v/v) ether. Hexane alone elutes hydrocarbons and squalene and with gradually increasing concentrations of ether, sterol esters, methyl esters, triacylglycerols, free fatty acids, cholesterol, diacylglycerols and monoacylglycerols are eluted, in that order.

Complex lipids (phospholipids, acylglycerols and cholesterol esters, for instance) contain mixtures of esterified fatty acids, both saturated and unsaturated. These may be released by saponification as described in Chapters 2 (Section 2.1.5) and 7 (Section 7.1.1). It is convenient to use methanolic NaOH so that the methyl esters of the fatty acids are produced. These can then be separated by TLC and GLC.

TABLE 5.1:   *The more common saturated fatty acids*

Chain length	Systematic name	Trivial name
12	Dodecanoic	Lauric
14	Tetradecanoic	Myristic
16	Hexadecanoic	Palmitic
18	Octadecanoic	Stearic
20	Eicosanoic	Arachidic

TABLE 5.2: The more common unsaturated fatty acids

Symbol	Systematic name	Trivial name
16:1	*cis*-9-Hexadecenoic	Palmitoleic
18:1	*cis*-9-Octadecenoic	Oleic
18:2	*cis*-9,12-Octadecadienoic	Linoleic
18:3	*cis*-6,9,12-Octadecatrienoic	γ-Linolenic (GLA)
20:4	*cis*-5,8,11,14-Eicosatetrasenoic	Arachidonic
20:5	*cis*-5,8,11,14,17-Eicosapentaenoic	EPA

TLC on silica gel impregnated with silver nitrate will separate the methyl esters of fatty acids according to their degree of unsaturation. Bands corresponding to saturated, monoenoic, dienoic and trienoic esters will appear on spraying with rhodamine 6G. The esters can be eluted from the silica with ether and then subjected to GLC.

5.1.3 GLC of fatty acids

Methyl esters of fatty acids for GLC are best prepared by treatment of the lipids in which they are combined (acylglycerols, phospholipids) with boiling methanolic HCl. After cooling, water is added and the methyl esters are extracted into light petroleum. Free fatty acids may also be methylated in this way. An alternative method uses diazomethane but this is both toxic and explosive and so is best avoided.

The methyl esters of fatty acids are readily separated by GLC, often on a stationary phase of polyethylene glycol. On such a column fatty acid esters of shorter chain length will elute first and for a particular chain length retention time increases with the number of double bonds. With nonpolar stationary phases such as Apiezon-L grease, the unsaturated esters will elute more readily than the saturated ones of the same chain length.

The stationary phases are held on supports of diatomaceous earth. With unknown mixtures of fatty acids it is best to carry out both polar and nonpolar GLC. The retention times of the unknowns relative to a standard of methyl palmitate are determined, enabling identification by comparison with published values such as those given by Kates [1].

Separations may be improved by temperature programing. For instance with a polyethylene glycol succinate stationary phase and nitrogen carrier gas, the temperature would be held at 185°C for 5 min and then raised to 225°C at a rate of 1° min^{-1}. FIDs are used in fatty acid separations (see Chapter 2, Section 2.2.2). Capillary columns usually give better separations than the normal type. An example is given in *Figure 5.2*.

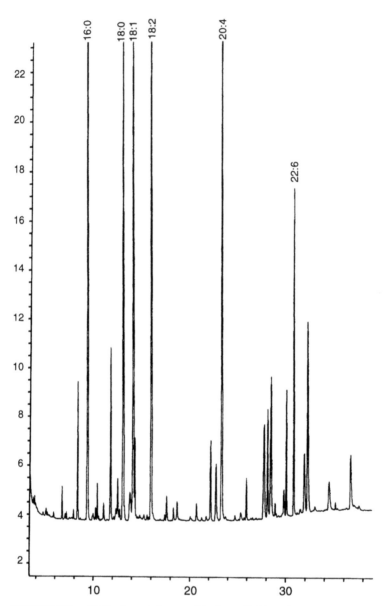

FIGURE 5.2: *GLC of total fatty acids from lipids of normal human erythrocytes. A 50 m capillary column was used with a CP-sil 88 polar stationary phase and temperature program as follows: 150° to 200° at 1° min^{-2}, then 2° min^{-1} to 240° and 6 min at 240°C. From the laboratory of Dr Nigel Lawson, Clinical Chemistry, City Hospital, Nottingham, UK.*

For a homologous series of fatty acid esters the change in retention time with chain length is constant, giving a CH_2-separation factor, F.

For two fatty acids of chain lengths n and $n+2$, having retention times t_n and t_{n+2} respectively, the following relationship holds:

$$\frac{t_{n+2}}{t_n} = F^2.$$

This formula can be used to identify saturated acids from their retention times if a standard of known chain length is used. The constant F will of course vary with temperature and carrier gas flow rate.

5.1.4 HPLC of fatty acids

Fatty acid methyl esters can be separated by HPLC. This can be either by absorption chromatography using silica with relatively nonpolar solvents, or as reversed phase HPLC where the column has long hydrocarbon chains and more polar solvents are used for elution. Classical on-line HPLC detection methods are not very satisfactory, however, so that GLC using FIDs is the method of choice.

5.2 Prostaglandins and leukotrienes (eicosanoids)

The prostanoids, comprising prostaglandins (PGs), prostacyclin and thromboxanes, are derivatives of the C_{20} cyclopentane compound prostanoic acid. The biosynthesis of the best-known 2-series begins with arachidonic acid, that of the 1-series begins with a $C_{20:3}$ acid and that of the 3-series with a $C_{20:5}$ acid. Prostanoids are rapidly destroyed in tissues so that they can only act as local hormones, having actions such as platelet aggregation and the contraction or relaxation of muscle. Thromboxane A_2 is particularly active (and labile), causing contraction of vascular smooth muscle and platelet aggregation at concentrations of 20 nM or less. Prostacyclin has opposite effects and will therefore cause dilation of blood vessels and inhibition of thrombus formation. It will be clear that these substances are important in relation to blood clotting and therefore to coronary thrombosis. Cyclo-oxygenase is the initial regulatory enzyme in their formation and is irreversibly inactivated by aspirin.

The enzyme lipoxygenase forms the leukotrienes from arachidonic acid in leukocytes, macrophages and other cells in response to immunological stimuli. Leukotrienes cause contraction of bronchial muscle, vascular permeability and activation of leukocytes, having a key role in inflammation. Some leukotriene and prostanoid structures are given in *Figure 5.3*.

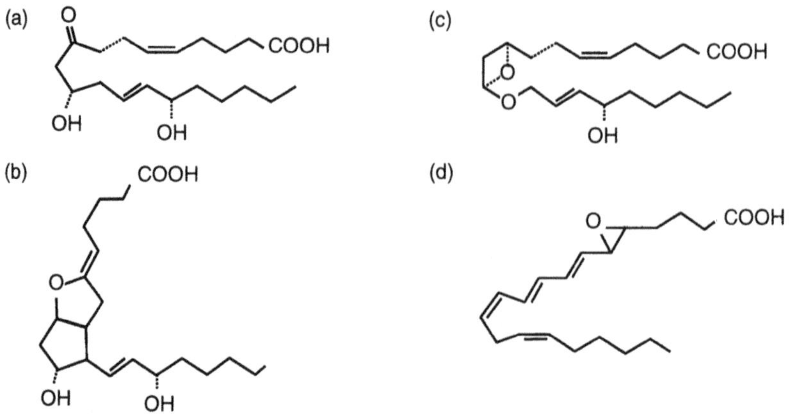

FIGURE 5.3: *Eicosanoid structures. (a) PGE$_2$; (b) prostacyclin; (c) thromboxane A$_2$; (d) leukotriene A$_4$.*

5.2.1 Analysis of eicosanoids

Most analyses will be of blood or samples of mammalian tissue and it is important to remember that levels of prostanoids can change rapidly after sampling, by biosynthesis or release from platelets. GLC with ECDs is able to detect as little as 1 ng. For example, tissue PGEs can be dehydrated to PGBs by treatment with methanolic KOH. Addition of aqueous HCl allows the PGBs to be extracted into ether. The carboxyl is then methylated with diazomethane and a trimethylsilyl derivative is prepared for GLC.

RIA can also be used. Antigens are made by coupling the prostanoid carboxyl to protein. The review of Samuelsson *et al.* [2] gives references dealing also with GC–MS. Commercial kits are available for the RIA of several prostanoids and leukotrienes, including thromboxane B$_2$, leukotrienes B$_4$, C$_4$ and D$_4$, prostaglandins D$_2$, E$_2$ and F$_2$ and 6-keto prostaglandin F$_1$. Some are assayed as breakdown products, the 6-keto PGF$_1$ coming from PGI$_2$ and thromboxane B$_2$ from thromboxane A$_2$.

Analytical methods for the leukotrienes include reversed phase HPLC, RIA and MS [3].

References

1. Kates, M. (1986) *Techniques of Lipidology* (2nd edn). Elsevier, Amsterdam.
2. Samuelsson, B., Goldyne M., Grandstrom, E., Hamberg, M., Hammarstrom, S. and Malmsten, C. (1978) *Ann. Rev. Biochem.*, **47**, 997–1029.
3. Hammarstrom, S. (1983) *Ann. Rev. Biochem.*, **52**, 355–377.

6 Esters

6.1 Acylglycerols

The older names for glycerol esterified with long chain fatty acids (*Figure 6.1*) were monoglyceride, diglyceride and triglyceride, i.e. glycerides. These are now known as monoacylglycerol, diacylglycerol and triacylglycerol respectively, i.e. acylglycerols.

Triacylglycerol is the most important storage fat, being found in animal depot fats, seed oils and milk. Natural triacylglycerols contain a variety of fatty acids and these are not randomly distributed among the 1-, 2- and 3-positions of the glycerol. Methods for determining the fatty acid composition at all positions are outlined in Section 2.6.2. Animal depot fats have saturated acids in position 1 and unsaturated or shorter chain acids in position 2. The population at position 3 is more random.

6.1.1 Separation of acylglycerols

A washed chloroform–methanol extract of a tissue such as liver will contain triacylglycerols and phospholipids together with free and esterified cholesterol. Smaller quantities of free fatty acids, monoacylglycerol and diacylglycerol may also be present. Such an extract is

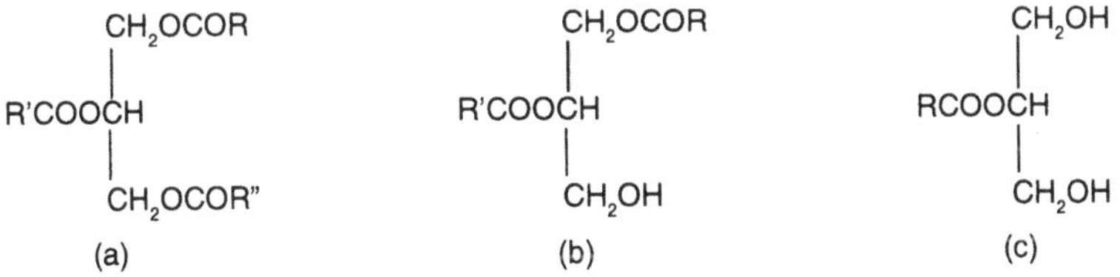

FIGURE 6.1: *Structures of acylglycerols. (a) Triacylglycerol; (b) sn-1,2-diacylglycerol; (c) 2-monoacylglycerol.*

best dried down in a rotary evaporator and the lipids redissolved in a small volume of chloroform for application to a column of silica. Continued elution with chloroform will remove acylglycerols, free fatty acids, cholesterol and cholesterol esters. The chloroform is evaporated *in vacuo* and the neutral lipid fraction redissolved in *n*-hexane so that the lipids can be separated on a column of Florisil (magnesium silicate) impregnated with boric acid to prevent isomerization of mono- and diacylglycerols [1]. Elution with hexane containing 5% (v/v) diethyl ether will remove cholesterol esters and triacylglycerols are then eluted with hexane containing 15% (v/v) diethyl ether. Increasing proportions of diethyl ether will then remove cholesterol, diacylglycerols and monoacylglycerols in turn. Free fatty acids require acidified diethyl ether for their removal from the column. This is an example of adsorption chromatography, the polar lipids being bound more strongly to the Florisil and thus requiring more polar solvents for their removal.

6.1.2 Analysis of acylglycerols

The contents of the various eluted fractions can be checked by TLC on silica gel impregnated with boric acid. The identity of the acylglycerols can be confirmed by IR spectroscopy and by chemical estimations. Fatty acid content is determined by saponification with ethanolic KOH, acidification and extraction with light petroleum for titration. This is performed by removing the solvent and redissolving the fatty acids in methanol–water (9:1 v/v) with a little ortho-cresol red as indicator. Titration is with methanolic 0.025 M NaOH (yellow to red end-point) and is performed hot. Ester bonds react with hydroxylamine and ferric perchlorate to give a purple complex which provides a colorimetric analysis [1]. Glycerol may be determined by the periodate–chromotropic acid colorimetric method. These methods should give the fatty acid:glycerol:ester molecular ratios for the various acyl glycerols (e.g. 2:1:2 for diacylglycerol).

6.1.3 Fatty acids of acylglycerols

The triacylglycerol fraction can be separated according to the degree of unsaturation of its components by TLC on silica gel impregnated with silver nitrate. A normal precoated plate is immersed in $AgNO_3$ solution (10% w/v in acetonitrile) for 30 min, dried and heated at 110°C for 1 h. Bands corresponding to saturated, monoenoic, dienoic, trienoic and tetraenoic acylglycerols can be obtained. More highly unsaturated species remain near the origin but can be separated using more polar solvents (*Figure 6.2*). The bands are detected by dichlorofluorescein and the triacylglycerols eluted from the plate with

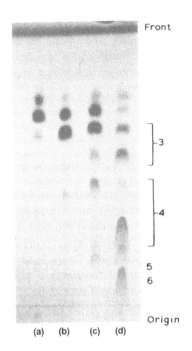

Front

3

4

5
6

Origin

(a) (b)   (c)   (d)

FIGURE 6.2: *Separation of triacylglycerol species by argentation TLC. Solvent: isopropanol–chloroform, 1.5:98.5 (v/v); spots located by spraying with 50% sulfuric acid and charring. (**a**) Palm oil; (**b**) olive oil; (**c**) groundnut oil; (**d**) cottonseed oil. The numbers represent the total number of double bonds in each triacylglycerol molecule. Reproduced from ref. 2 (Figure 4.4) with permission from the authors and Chapman and Hall.*

chloroform–methanol (9:1 v/v) for GLC, which will separate them on the basis of fatty acid chain length, unsaturation having little effect. Separation of triacylglycerols differing in molecular weight by only two carbons is easily possible. A relatively short column is used, with a nonpolar stationary phase such as 1–3% SE-30 on Chromosorb W, with temperature programing from 175 to 350°C (1–4° min⁻¹). References are given in ref. 1.

Natural triacylglycerol mixtures can also be separated by reversed phase HPLC on microspheres of silica gel to which long chain aliphatic groups are covalently bound [1]. Stereospecific analysis of triacylglycerols, i.e. identification of fatty acids on each of the three glycerol hydroxyls, can be carried out as described in Section 2.6.2.

Monoacylglycerols and diacylglycerols can also be separated according to their degree of unsaturation by the same argentation TLC, but they must first be acetylated in pyridine–acetic anhydride. As with the triacylglycerols, the resulting sub-fractions are then analyzed by GLC to identify the molecular species present. Direct molecular species analysis of diacylglycerols is possible by reversed phase HPLC as above, but the *tert*-butyl dimethylsilyl derivative of the free hydroxyl must first be made. This provides a way of analyzing the molecular species of phospholipids from which the diacylglycerol residue has been removed by phospholipase C action. The fatty acid composition of the diacylglycerol released from the phosphoinositides as a result of receptor-mediated phospholipase C hydrolysis is of particular interest, since the diacylglycerol acts as a second messenger to stimulate protein kinase C. The major diacylglycerol component is *sn*-1-stearoyl, 2-arachidonoyl glycerol.

6.2 Wax esters

These are esters of general structure RCOOR', where both R and R' are long aliphatic chains. They occur in nature in some bacteria, certain seed oils, the oil of the sperm whale and some deep sea fishes. The greatest amount globally is found in the zooplankton of the oceans. Cutin and suberin are more polar polymeric compounds found on the outer surfaces of plants. Cutin contains dihydroxy fatty acids and suberin has dicarboxylic fatty acids and ω-hydroxy fatty acids.

Identification of wax esters [1] requires a preliminary fractionation on a silica column from which they can be eluted with benzene. A pure wax monoester fraction is obtained by preparative TLC in benzene and its identity can be checked by IR or NMR spectroscopy. Individual species can be separated by preparative reversed phase HPLC and each one saponified for the identification of the resulting long chain acid and alcohol by GLC. The free long chain alcohols are separable in this way, but the acids need to be converted to their methyl esters.

6.3 Alkylglycerols

These compounds contain hydrocarbon chains linked to glycerol by an ether rather than an ester linkage. The ether linkage is illustrated in Chapter 7 (*Figure 7.2b*) along with the unsaturated ether (plasmalogen) group (*Figure 7.2a*). Marine organisms, mammals and birds all contain small amounts of monoalkyl ethers of glycerol, the hydrocarbon residue being linked to the *sn*-1 position. The commonest such compounds are chimyl alcohol (*sn*-1-*O*-hexadecyl glycerol), batyl alcohol (*sn*-1-*O*-octadecyl glycerol) and selachyl alcohol (*sn*-1-*O*-octadec-*cis*-9-enyl glycerol). Compounds of this type also occur in the acylated form (*sn*-1-alkyl 2,3-diacylglycerol). The so-called 'neutral plasmalogens' are lipids of this type, except that the 1-position has an unsaturated ether (alk-1'-enyl group).

6.3.1 Analytical methods for alkylglycerols

The 1-alkyl 2,3-diacylglycerols and neutral plasmalogens will appear in the triacylglycerol fraction after chromatography on a silicic acid column. Mild alkaline hydrolysis will remove the acyl groups to leave monoalkyl ethers which can be purified by preparative TLC. A mixture of alk-1-enyl- and alkylglycerols prepared in this way can be analyzed by GLC after acetylation followed by mild acid hydrolysis, which produces alkyldiacetylglycerols and free aldehydes [1]. The acetylation is done with acetic anhydride in dry pyridine. After the addition of methanol and water the acetates are extracted into light petroleum.

Alkylglycerols eluted with diethyl ether from TLC plates can be identified from their infrared spectra. There should be strong ester and ether bands and for the neutral plasmalogens characteristic *cis*-vinyl ether bands. NMR spectrometry also reveals characteristic signals for CH_2 groups adjoining the ether linkage.

References

1. Kates, M. (1986) *Techniques of Lipidology* (2nd edn). Elsevier, Amsterdam.
2. Gurr, M.I. and Harwood, J.L. (1991) *Lipid Biochemistry: an Introduction* (4th edn). Chapman and Hall, London.

7 Phospholipids, Sulfolipids and Related Compounds

7.1 Phospholipids with amino-groups

This category includes phosphatidylcholine (lecithin), the most abundant phospholipid, together with phosphatidylethanolamine (cephalin) and phosphatidylserine. Plant tissues contain *N*-acyl derivatives of phosphatidylethanolamine, the acyl groups being derived from a long chain fatty acid such as stearic acid [1]. These compounds all have a glycerol backbone and a phosphate group in the *sn*-3 position. Positions *sn*-1 and *sn*-2 of the glycerol are usually linked to long chain acyl groups. As well as these diacyl phospholipids, there are smaller amounts of *sn*-1-alk-1'enyl, *sn*-2-acyl and *sn*-1-alkyl, *sn*-2-acyl compounds in animal tissues, the former being known as plasmalogens. Structures of these compounds are given in *Figures 7.1* and *7.2*.

A second group of amino-phospholipids has the long chain base sphingosine instead of glycerol. The best-known phospho-sphingolipid is sphingomyelin, a phosphorylcholine derivative of *N*-acyl-sphingosine (ceramide). The analogous phosphorylethanolamine derivative (ceramide phosphorylethanolamine) has also been shown to exist in shellfish, insects, protozoa and some anaerobic bacteria [1]. In plants, sphingolipids containing inositol and various sugars have been described (phytoglycolipids). A compound of this type has recently been characterized in protozoa which infect the genitourinary tract [2]. It has a ceramide residue linked to inositol 1-phosphate. In turn, the inositol 4-position is attached to fucose and its 3-position to phosphorylethanolamine.

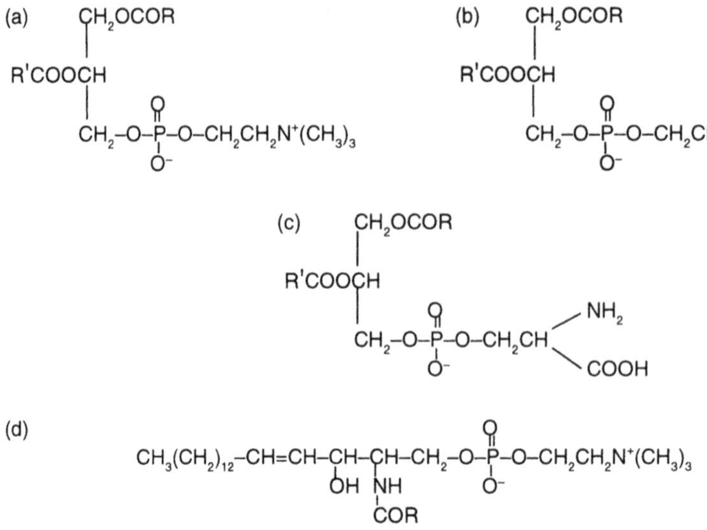

(a)

$$CH_2OCOR$$
$$R'COOCH$$
$$CH_2-O-\overset{\overset{O}{\|}}{\underset{\underset{O^-}{|}}{P}}-O-CH_2CH_2N^+(CH_3)_3$$

(b)

$$CH_2OCOR$$
$$R'COOCH$$
$$CH_2-O-\overset{\overset{O}{\|}}{\underset{\underset{O^-}{|}}{P}}-O-CH_2CH_2NH_2$$

(c)

$$CH_2OCOR$$
$$R'COOCH$$
$$CH_2-O-\overset{\overset{O}{\|}}{\underset{\underset{O^-}{|}}{P}}-O-CH_2CH\overset{NH_2}{\underset{COOH}{<}}$$

(d)

$$CH_3(CH_2)_{12}-CH=CH-\underset{\underset{OH}{|}}{CH}-\underset{\underset{\underset{COR}{|}}{NH}}{CH}-CH_2-O-\overset{\overset{O}{\|}}{\underset{\underset{O^-}{|}}{P}}-O-CH_2CH_2N^+(CH_3)_3$$

FIGURE 7.1: *Structures of phospholipids with amino-groups:*
(a) phosphatidylcholine; (b) phosphatidylethanolamine;
(c) phosphatidylserine; (d) sphingomyelin.

7.1.1 Hydrolytic procedures in analysis

Mild alkaline hydrolysis (0.1 M NaOH in methanol at 37°C) will remove the fatty acids from diacyl phospholipids, though alkyl and alk-1'-enyl residues are not hydrolyzed. Phospho-sphingolipids are also resistant to mild alkaline treatment, so this provides a way of isolating them from tissues. Following alkaline hydrolysis phospholipids are converted to water-soluble esters such as glycerophosphocholine and the methyl esters of fatty acids, both of which are easily separated from sphingomyelin. The phosphate esters can be removed into the aqueous phase by adding chloroform and 0.2 M HCl to acidify the hydrolyzate. After a further wash with water the chloroform layer is treated with methanol and aqueous 0.05 M HCl containing 25 mM HgCl$_2$ to remove the alkenyl group from lysoplasmalogens. The chloroform layer is then dried down *in vacuo* and

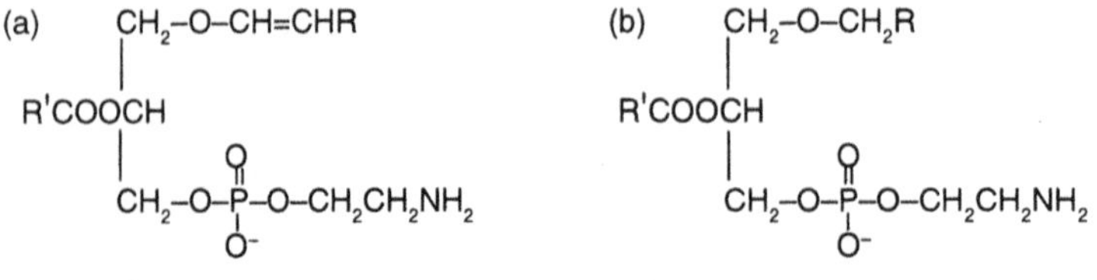

(a)

$$CH_2-O-CH=CHR$$
$$R'COOCH$$
$$CH_2-O-\overset{\overset{O}{\|}}{\underset{\underset{O^-}{|}}{P}}-O-CH_2CH_2NH_2$$

(b)

$$CH_2-O-CH_2R$$
$$R'COOCH$$
$$CH_2-O-\overset{\overset{O}{\|}}{\underset{\underset{O^-}{|}}{P}}-O-CH_2CH_2NH_2$$

FIGURE 7.2: *(a) Ethanolamine plasmalogen (1-alk-1'-enyl,2-acyl compound); (b) alkyl ether form of ethanolamine phospholipid (1-alkyl,2-acyl compound).*

redissolved in chloroform–methanol (8:1 v/v) for application to a silicic acid column. Methyl esters and any cerebrosides (particularly if the starting material is brain tissue; see Chapter 8) are removed by chloroform–methanol (4:1 v/v) and sphingomyelin by chloro-form–methanol (1:4 v/v).

Selective hydrolysis can be used in combination with the colorimetric analysis of organic phosphates by the phosphomolybdate method (Section 7.1.2) to fractionate phospholipids. Mild alkaline hydrolysis and phase separation by adding chloroform gives an aqueous phase containing phosphate esters from all the diacyl phospholipids. The organic phase is then treated with $HCl/HgCl_2$ and water addition gives an aqueous phase with phosphate esters from plasmalogens and an organic phase of which the phosphate content comes from sphingomyelin and any 1-alkyl phospholipids. Phosphate deter-minations on the various phases give: (i) total diacyl phospholipid, (ii) plasmalogens; and (iii) sphingomyelin plus alkyl ether phospholipids.

After mild alkaline hydrolysis, the phosphate esters of the aqueous phase can be separated by two-dimensional paper chromatography [1,3] to give the total phospholipid pattern of a tissue. Spots corresponding to the various phospholipids can be eluted and their phosphate content estimated. This method is less used nowadays, since direct TLC separation of the intact phospholipids provides a more convenient analysis by phosphate content (see Section 7.1.3).

7.1.2 Estimation of phosphate

The best colorimetric method depends on the formation of a blue complex on the reduction of phosphomolybdate. The lipid sample, containing 0.5–5 µg of phosphorus, is first digested with 72% (v/v) perchloric acid to convert phosphate esters to inorganic orthophos-phate. The minimum amount of lipid should be used since explosions can take place with too much organic material. For this reason, solvents such as chloroform should be evaporated before adding the perchloric acid. The method of Bartlett [4] is then followed. This involves addition of ammonium molybdate solution, followed by the reducing agent, a sodium bisulfite solution containing 1-amino-2-napthol-4-sulfonic acid. The mixture is then heated for 7 min at 100°C and the optical density is read at 830 nm. Standards of KH_2PO_4 may be used.

7.1.3 Separation of phospholipids

TLC is the method of choice for the analysis of tissue phospholipids. Two-dimensional TLC will separate not only the amino-phospholipids

TABLE 7.1:   *Two-dimensional TLC of phospholipids*[a]

	R_f values	
	Solvent 1	Solvent 2
Phosphatidylcholine	0.39	0.60
Phosphatidylethanolamine	0.59	0.62
Lysophosphatidylcholine	0.24	0.34
Phosphatidylserine	0.33	0.49
Phosphatidylinositol	0.33	0.17
Phosphatidic acid	0.59	0.49
Cardiolipin	0.69	0.45
Phosphatidylglycerol	0.47	0.42
Lysophosphatidylethanolamine	0.45	0.32

[a] Silica gel plates were developed in solvent 1: chloroform–methanol–acetic acid–water (50:20:7:3 by vol.); and then in solvent 2: chloroform–methanol–40% aqueous methylamine–water. (13:7:1:1 by vol.).

but also the acidic ones discussed in the following sections. *Table 7.1* gives the R_f values of phospholipids separated in this way. Another example is given in *Figures 7.3* and *7.4*. Glass plates (20 cm square) were coated with silica gel in 3% (w/v) magnesium acetate and, after drying at 110°, lines were scored to divide them into four quarters (*Figure 7.3*). The lipid extracts from adrenal chromaffin granules were then applied at each corner. The photograph is of a plate stained with iodine vapor, but for more specific identifications Dragendorff reagent will identify choline lipids and ninhydrin gives a purple color with ethanolamine and serine derivatives [3]. The Dragendorff reagent is a solution of bismuth nitrate in acetic acid which is mixed with potassium iodide solution just before use. Choline-containing lipids give an orange color without heating. The ninhydrin spray (0.2% w/v in acetone) produces purple spots after a few hours at room temperature. The four sections of the plate can be stained with different reagents, e.g. iodine, phosphate spray, ninhydrin and Dragendorff reagent. Other examples of phospholipid separations by TLC are given by Kates [3], and tables of R_f values in different solvents are given in ref. 1. For quantitative analysis, the spots obtained with iodine are outlined, the iodine allowed to sublime away and the silica of the spot scraped off for phosphate determination.

7.2 Phosphoinositides

The inositol-containing phospholipids, phosphatidylinositol, phosphatidylinositol 4-phosphate and phosphatidylinositol 4,5-bisphosphate

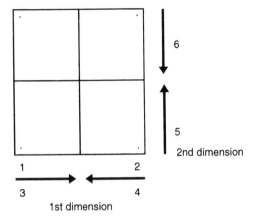

FIGURE 7.3: *Two-dimensional TLC method for phospholipid separation. The method gives four separations on one plate. The first solvent is run twice. Solvent 1: chloroform–methanol–ammonia (65:25:5 by vol.); solvent 2: chloroform–acetone–methanol–glacial acetic acid–water (30:40:10:10:5 by vol.).*

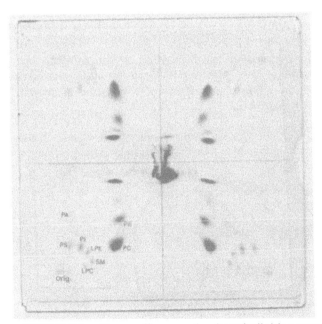

FIGURE 7.4: *Chromaffin granule phospholipids separated as in Figure 7.3. Orig, origin, PA, phosphatidic acid; PE, phosphatidylethanolamine; PS, phosphatidylserine; PI, phosphatidylinositol; PC, phosphatidylcholine; LPE, lysophosphatidylethanolamine; SM, sphingomyelin; LPC, lysophosphatidylcholine. Dark spots at the solvent front are due to cholesterol and glycerides. Reproduced from Nor Azila Mohd. Adnan (1981), PhD Thesis, University of Nottingham with permission from the author.*

are now known to be of considerable importance in relation to receptors and second messengers. Activation of various cell surface receptors leads via a G-protein to phospholipase C hydrolysis of phosphatidylinositol 4,5-bisphosphate, releasing inositol 1,4,5-trisphosphate and diacylglycerol. Both products can act as second messengers, the former releasing intracellular Ca^{2+} and the latter activating protein kinase C. Two types of analysis are of particular interest; firstly, the separation of the various intact phosphoinositides and secondly, labeling tissues with [^{3}H]inositol and separating the various radioactive inositol phosphates released upon receptor activation. Before going on to the analytical methods, it is necessary to outline the metabolism of the phosphoinositides and their hydrolysis products.

7.2.1 Phosphoinositide and inositol phosphate metabolism

This is outlined in *Figure 7.5,* and *Figure 7.6* clarifies the inositol ring numbering. Phosphatidylinositol is the parent compound which is phosphorylated by two ATP-requiring kinases to phosphatidylinositol 4-phosphate and the 4,5-bisphosphate. Another kinase can phosphorylate all three lipids at the 3-position of the inositol, but much smaller amounts of the 3-phosphorylated phosphoinositides occur in tissues. They may be important in the regulation of growth rather than in signal transduction.

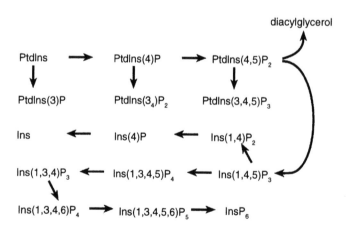

FIGURE 7.5: *Pathways of phosphoinositide metabolism in mammalian tissues. PtdIns, phosphatidylinositol; Ins, myo-inositol. Numbers refer to inositol ring hydroxyls.*

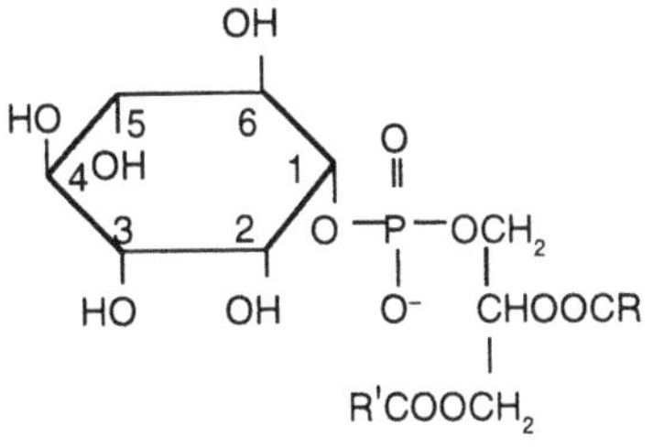

FIGURE 7.6: *Phosphatidylinositol, showing ring numbering of the 1D-inositol 1-phosphate structure.*

Receptor activation leads to hydrolysis of phosphatidylinositol 4,5-bisphosphate by a phospholipase C and the resulting inositol 1,4,5-trisphosphate can be metabolized in two ways. In the first, phosphatases hydrolyze it sequentially to free inositol and in the second a kinase adds a further phosphate to the 3-hydroxyl. The resulting inositol 1,3,4,5-tetrakisphosphate is then hydrolyzed to the 1,3,4-compound, which in turn can be phosphorylated to the hexakisphosphate, phytic acid, or hydrolyzed to lower phosphates and finally to free inositol.

7.2.2 Separation of the phosphoinositides

The polyphosphoinositides (a collective term for the phosphorylated derivatives of phosphatidylinositol) are so polar that they are not extracted from tissues by neutral solvents such as chloroform–methanol mixtures. An acidified solvent such as chloroform–methanol–conc. HCl (200:100:1 by vol.) is required.

TLC is the method of choice for separating the phosphoinositides from other phospholipids. Silicic acid plates are used and are best impregnated with 1% (w/v) potassium oxalate–2 mM EDTA [5]. Suitable solvents, for example chloroform–methanol–4 M ammonia (9:7:2 by vol.), will separate the phosphoinositides from other phospholipids which run more quickly, and from each other. Phosphatidylinositol is not easy to separate from phosphatidylserine unless a two-dimensional method is used, with for instance, chloroform–methanol–glacial acetic acid–water (80:40:7.4:1.2 by vol.) as the second solvent. Methods are also available for separation of the 3-phosphorylated phosphoinositides [5].

7.2.3 Receptor-linked phosphoinositide changes

The simplest and most widely used method of studying receptor-mediated phosphoinositide metabolism involves radiolabeling the phosphoinositide pool with [³H]inositol, preferably to radioisotopic equilibrium, and then following the generation of [³H]inositol phosphates after addition of a suitable agonist at a physiological concentration. The incubation medium usually includes 5 mM LiCl to inhibit inositol monophosphatase so that total inositol phosphate accumulation provides a better measure of phosphatidylinositol 4,5-bisphosphate hydrolysis. A brief period of exposure to the agonist is recommended, 5 min or less. Using ion-exchange resin columns, the total inositol phosphates can be separated from free inositol and their labeling measured. Individual inositol phosphates can be separated by HPLC. Methods are also available for the mass determination of inositol 1,4,5-trisphosphate using its specific binding protein. Though not often used, the assay of diacylglycerol released by receptor activation has been developed. These methods are outlined in the next two sections.

7.2.4 Separation and estimation of inositol phosphates

These procedures are used in studies of the phosphoinositide response to the activation of receptors. After labeling to equilibrium *in vivo* or *in vitro* with [³H]inositol, slices of the tissue are incubated for a few minutes with the agonist, for example the acetylcholine derivative carbachol, which activates muscarinic receptors, and the incubation is terminated by the addition of trichloroacetic acid. Details are given in ref. 5. After removal of the acid by ether extraction, the aqueous layer is applied to a small column of the anion-exchange resin Dowex AG1 and the inositol phosphates are eluted, for instance, with increasing concentrations of ammonium formate in formic acid. Fractions are collected for the counting of ³H with an appropriate scintillant. A typical separation is shown in *Figure 7.7* and can be compared with the peaks obtained from labeled tissue not treated with the agonist. Increased labeling due to receptor activation is seen in all peaks except glycerophosphorylinositol, inositol pentakisphosphate and phytic acid.

For separation of the isomeric inositol phosphates, HPLC is the method of choice [5]. Columns of Partisil SAX are commonly used with gradient elution by acid–salt mixtures. The samples may be 'spiked' with nucleotides so that an on-line UV monitor measures column performance. *Figure 7.8* shows how nucleotide pairs (AMP and GMP, ADP and GDP, ATP and GTP) are used in HPLC separations of the inositol phosphates produced by muscarinic stimulation of brain slices. A number of radiolabeled standards including inositol 1-phosphate, inositol 4-phosphate, inositol 1,4-bisphosphate, the 1,3,4- and 1,4,5-trisphosphates and inositol 1,3,4,5-tetrakisphosphate are available commercially.

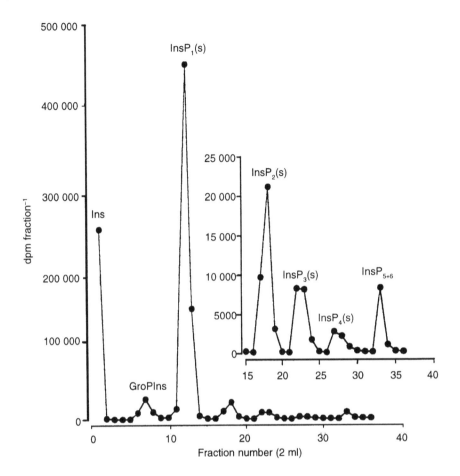

FIGURE 7.7: *Dowex AG1 × 8 chromatography (1 ml resin in 0.6 cm diameter column) of inositol phosphates produced by muscarinic stimulation of astrocytoma cells labeled with [³H]inositol for 2–3 days. The inset shows an expanded scale for the fractions indicated. Eluants are as follows: free inositol, water; GroPIns, 60 mM ammonium formate (AF); InsP₁, 200 mM AF; InsP₂, 500 mM AF, 100 mM formic acid (F); InsP₃, 800 mM AF, 100 mM F; InsP₄, 1.0 M AF, 100 mM F; InsP₅ plus InsP₆, 2.0 M AF, 100 mM F. (s) indicates more than one isomer.*

For the mass determination of inositol 1,4,5-trisphosphate a radio-ligand binding assay has been developed [5]. This makes use of a specific high-affinity inositol 1,4,5-trisphosphate-binding protein from the microsomal fraction of bovine adrenal cortex. Accurate estimations of concentrations as low as 0.5 nM may be made.

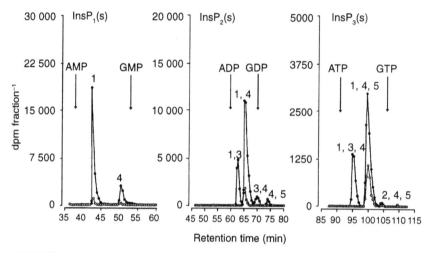

FIGURE 7.8: *Analysis of inositol phosphate isomers by HPLC. [³H]inositol-labeled brain slices were incubated for 5 min (InsP₁) or 30 min (InsP₂ and InsP₃) with 1 mM carbachol and 10 mM LiCl. Open circles indicate LiCl only. Arrows show the approximate retention time of the nucleotides used in 'spiking'.*

7.2.5 Estimation of diacylglycerol

It is sometimes of interest to determine changes in the concentration of diacylglycerol in response to receptor-activated hydrolysis of phosphoinositides. Methods developed by Zhu and Eichberg are given in a recent paper [6]. The stimulated tissue is extracted with chloroform–methanol and the diacylglycerol is purified by TLC using borate-impregnated silica gel plates. The diacylglycerol spot is cut out and extracted with chloroform–methanol (1:1 v/v). The extract is washed with 1 M NaCl, giving a lower phase containing the diacylglycerol. Using diacylglycerol kinase from *Escherichia coli* and [γ-³²P]-ATP, the purified diacylglycerol is converted to [³²P]phosphatidic acid, which is separated by TLC and counted. Diacylglycerol mass is calculated from a standard curve generated using 50–1000 pmol of pure *sn*-1,2-diolein in the same system.

7.3 Phosphatidylinositol glycans

Glycosylated phosphatidylinositols, or PI glycans for short, were discovered as a result of the observation by Low [7] that PI-specific phospholipase C caused the release of certain proteins from cell

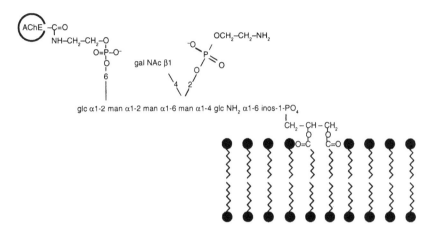

FIGURE 7.9: *The phosphatidylinositol glycan linked acetylcholinesterase (AChE) in the membrane of the electric organ in the electric fish* Torpedo marmorata.

membranes. These proteins are anchored to the membrane by covalently binding to PI. The C-terminal of the protein is linked to the ethanolamine amino-group of this glycan:

$$\text{ethanolamine–phosphate–(mannose)}_3\text{–glucosamine}$$

and the glucosamine reducing group is in turn linked to the 6-hydroxyl of PI.

Such PI glycans are found in the cell surfaces of protozoa (e.g. African trypanosomes), yeast, slime molds, fish and mammals. They contain the minimal glycan structure above, but this is modified in different ways by the addition of further sugars, ethanolamine phosphate residues and fatty acids [8]. An example is given in *Figure 7.9*. Some protozoa produce PI glycans in the glycolipid form, i.e. unattached to protein.

7.3.1 Detection of PI glycans

A phospholipase C which only hydrolyzes the glycosylated form of PI, and not the lipid itself or other phospholipids, has been isolated from *Trypanosoma brucei*. Treatment of cells or tissues with this enzyme releases the PI glycan-anchored protein with its glycan component attached. Specific antibodies are then used in immunoprecipitation and Western blotting. Confirmation may be obtained by labeling the original cells with [³H]ethanolamine, which will appear in the glycan residue.

Chemical and enzymic methods for the analysis of PI glycans are discussed in a review by Ferguson [9]. In the same volume there is a useful article by Low [10] on the phospholipases hydrolyzing these compounds. Both reviews give practical details.

7.4 Phosphonolipids

The first phosphonolipid was isolated from a sea anemone in 1963 [11] and such lipids also occur in shellfish, snails and protozoa. They contain the unusual C–P bond, in the form of aminoethylphosphonate. *Figure 7.10* shows the glycerophosphonolipid phosphatidylethylamine. There are also sphingophosphonolipids, one being the analog of sphingomyelin, whilst another possesses a galactose residue.

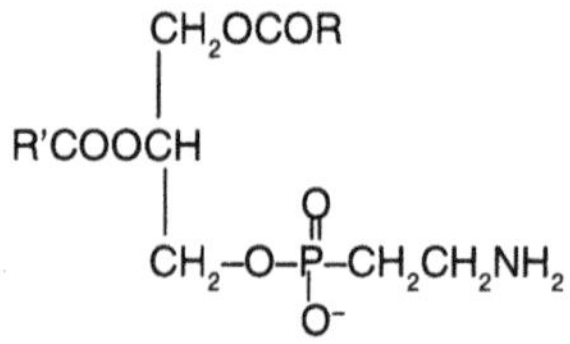

CH$_2$OCOR

R'COOCH

CH$_2$–O–P–CH$_2$CH$_2$NH$_2$

O$^-$

FIGURE 7.10: *3*-sn-*Phosphatidylethylamine, a glycerophosphonolipid.*

7.4.1 Separation and analysis of phosphonolipids

Where phosphonolipids are suspected to occur, they may be estimated in total as the difference between the total lipid phosphate of the sample (estimation of inorganic phosphate after digestion with perchloric acid) and the acid-hydrolyzable phosphate (inorganic phosphate liberated by hydrolysis in a sealed tube with 2 M HCl at 125°C). The C–P bond is stable to the 2 M HCl hydrolysis. Details are given by Kates [3].

Phosphonolipids of all types are extracted from tissues by the usual chloroform–methanol solvents. They may be separated from other lipids by chromatography on DEAE cellulose columns using chloroform–methanol mixtures [11] or by preparative two-dimensional TLC.

7.5 Alkylether phospholipids

In most tissues the diacyl forms of the phospholipids predominate, but small amounts of ether derivatives are also found. Phosphatidylethanolamine, phosphatidylcholine and phosphatidylserine all occur as plasmalogens (1-alk-1'-enyl,2-acyl compounds, *Figure 7.2*). The two former phospholipids are also found as 1-alkyl,2-acyl derivatives and in halophilic bacteria phosphatidylglycerol occurs as the diphytanyl ether (phytanic acid is 3,7,11,15-tetramethylhexadecanoic acid) (see also Section 4.1.2). Platelet-activating factor (*Figure 7.11*) is able at very low concentrations (5×10^{-11} M) to cause aggregation of blood platelets. It is the 1-alkyl,2-acetyl form of phosphatidylcholine.

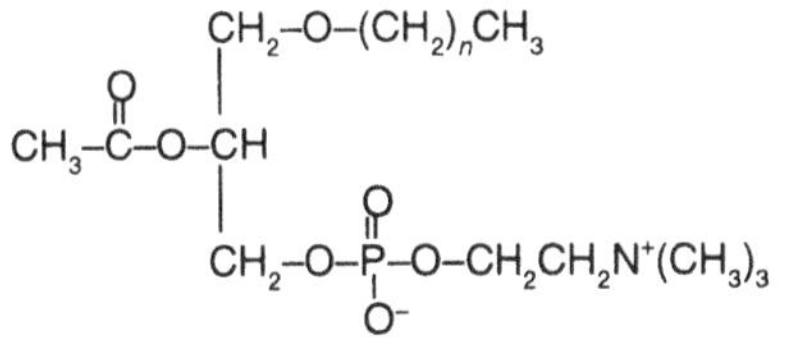

FIGURE 7.11: Platelet-activating factor.

7.5.1 Analysis of alkylether phospholipids

Hydrolytic methods for the analysis of plasmalogens and alkylether phospholipids are given in Section 7.1.1.

The diacyl-, alkylether and plasmalogen forms of a particular phospholipid (e.g. phosphatidylethanolamine) are not readily separable by chromatography. The three forms of this particular phospholipid, however, may be separated by TLC after dinitrophenylation followed by methylation with diazomethane [3].

An alternative analytical method is to treat the phospholipid with phospholipase C, thus producing diacyl-, alkylacyl- and alkenyl-acylglycerols. These may then be analyzed by HPLC as their acetate derivatives.

7.6 Sulfolipids

A number of sulfur-containing lipids are known, differing widely in chemical structure. Perhaps the most unusual is phosphatidyl-sulfocholine, the major polar lipid of the marine diatom *Nitzchia alba*. In this compound the nitrogen atom of choline is replaced by sulfur, with the loss of a methyl group [$OHCH_2CH_2S^+(CH_3)_2$]. The remaining sulfolipids can be divided into two groups, the sulfate derivatives and the sulfonolipids in which a carbon–sulfur bond is found.

7.6.1 Lipid sulfates

Sulfate esters of sterols occur in mammalian cells, e.g. cholesterol 3-*O*-sulfate, and in lower forms of life. The best known lipid sulfates are the sulfatides, found in myelin of nerve tissue. These are cerebroside sulfates, i.e. ceramide galactose-3-sulfate.

Long chain diol sulfates also containing chlorine or bromine have been found in a phytoflagellate, a sulfated triglycosyl diphytanyl ether in halophylic bacteria and sulfate esters of long chain acylated trehalose in tubercle bacilli [3]. Finally, a 6-sulfogalactosyldiacylglycerol has been found in various mammalian tissues.

7.6.2 Sulfonolipids

Plant chloroplast membranes contain sulfoquinovosyldiacylglycerol, quinovose being 6-deoxy-glucose. The structure is shown in *Figure 7.12;* see also Section 8.3. Mostly saturated acids, e.g. palmitic, are attached to the glycerol. Sphingolipids containing a sulfonic acid residue have been found in marine diatoms and certain bacteria [3].

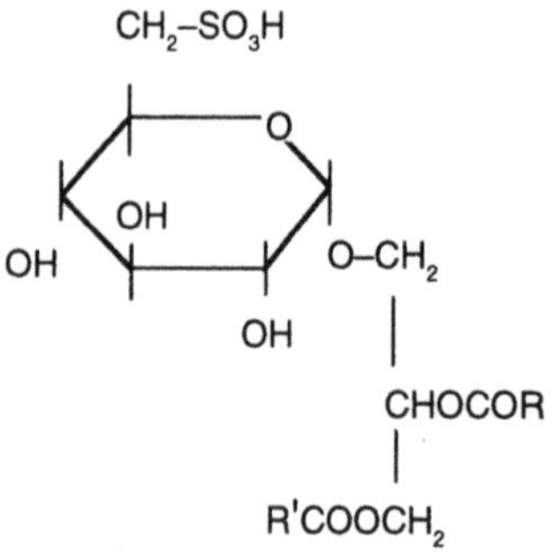

FIGURE 7.12: *The plant sulfonolipid sulfoquinovosyl-diacylglycerol.*

7.6.3 Analytical methods

Methods for the identification and determination of all these diverse compounds are not available. If an organism or tissue is suspected to contain a sulfolipid, a useful first step is to label it with [^{35}S]sulfate. Extraction with chloroform–methanol followed by TLC and auto-radiography will show up any sulfolipids. Various spray reagents detect sugar-containing sulfolipids [3] and there is also a spectro-photometric assay for cerebroside sulfate, based on formation of a colored complex with the cationic dye azure A [3].

References

1. Ansell, G.B., Hawthorne. J.N. and Dawson, R.M.C. (1973) *Form and Function of Phospholipids*. Elsevier, Amsterdam.
2. Costello, C.E., Glushka, J., van Halbeck, H. and Singh, B.N. (1993) *Glycobiology*, **3**, 261–269.
3. Kates, M. (1986) *Techniques of Lipidology* (2nd edn). Elsevier, Amsterdam.
4. Bartlett, G.R. (1959) *J. Biol. Chem.*, **234**, 466–468.
5. Batty, I.H., Carter, A.N., Challis, R.A.J. and Hawthorne, J.N. (1995) *Neurochemistry: a Practical Approach* (A.J. Turner and H.S. Bachelard, eds). Oxford University Press, Oxford.
6. Yorek, M.A., Dunlap, J.A., Stefani, M.R., Davidson, E.P., Zhu, X. and Eichberg, J. (1994) *J. Neurochem.*, **62**, 147–158.
7. Low, M.G. (1988) *Biochim. Biophys. Acta*, **988**, 427–454.
8. McConville, M.J. and Ferguson, M.A.J. (1993) *Biochem. J.*, **294**, 305–324.
9. Ferguson,M.A.J. (1992) in *Lipid Modification of Proteins: a Practical Approach* (N.M. Hooper and A.J. Turner, eds). Oxford University Press, Oxford, pp. 191–230.
10. Low, M.G. (1992) in *Lipid Modification of Proteins: a Practical Approach* (N.M. Hooper and A.J. Turner, eds). Oxford University Press, Oxford, pp. 117–154.
11. Hori, T., Horiguchi, M. and Hayashi, A. (1984) *Biochemistry of Natural C–P Compounds*. Shiga University, Japan.

8   Glycolipids

The broad definition of glycolipids as compounds containing carbohydrate covalently linked to a lipid-soluble moiety encompasses a wide range of substances. This broad group can be divided conveniently into four subgroups: glycosphingolipids, polyisoprenyl phosphosugars, glycosylglycerides and lipoteichoic acids.

8.1 Glycosphingolipids

8.1.1 Biological significance

These compounds are found as components of the outer leaflet of the plasma membrane of animal cells. A hydrophobic ceramide portion anchors a hydrophilic oligosaccharide chain to the surface of the cell. The expression of these compounds is often specific to a particular cell type or developmental stage. Their location and diversity of structure makes them well suited to a role in cell surface recognition. Examples of this function can be found in leukocyte adhesion to vascular endothelium, which involves the binding of proteins such as ELAM-1 (E-selectin) to carbohydrate determinants of glycosphingolipids [1]. Also important is the involvement of glycosphingolipids in neural cell–cell and cell–substratum recognition and adhesion.

Glycosphingolipids have also been reported (see e.g. ref. 1) to influence the activity of proteins in the same plasma membrane (*cis*-regulation) and in apposing membranes (*trans*-regulation). The proteins concerned are generally cell-surface receptors. It is thought that this may allow modulation of cell growth and differentiation, especially in developing nerve tissue.

The presence of glycosphingolipids as recognition and binding molecules at the cell surface provides several pathogenic organisms with convenient targets when invading cells. Several viruses have been reported to use host glycosphingolipids as adherence sites on host cell surfaces. This sort of phenomenon has also been implicated in the initiation of infection of host cells by bacteria. In some cases, for example cholera toxin, the protein exotocin released by bacteria gains entry to the target cell by first binding to a specific glycosphingolipid (ganglioside GM$_1$) on its surface.

8.1.2 Structures

In glycosphingolipids the lipid-soluble moiety is *N*-acylsphingosine (ceramide, *Figure 8.1*). A glucose or galactose unit is linked glycosidically to the primary hydroxyl group of this lipid. The monosaccharides shown in *Table 8.1* are attached at the reducing terminus of the glucose or galactose to form a chain of monosaccharide units constituting the glycan (oligosaccharide) part of the glycosphingolipid. The nature of the glycan is very variable (almost 300 different structures having been described) and influences function as well as the selection of appropriate isolation and analytical procedures. Those with neutral glycans of five or fewer monosaccharide units are the least polar and can be isolated efficiently using classical lipid solvents. Longer glycan chains render the molecule more hydrophilic, such that the use of more polar solvents may be advantageous. Anionic glycan chains, for example those characterizing gangliosides, all of which contain one or more residues of sialic acid, and those that carry sulfate residues may also cause the glycosphingolipid to dissolve more readily in polar solvents at alkaline or neutral pH.

Several attempts have been made to simplify and systematize the nomenclature of glycosphingolipids. The many complex glycan structures and the plethora of trivial names may sometimes appear

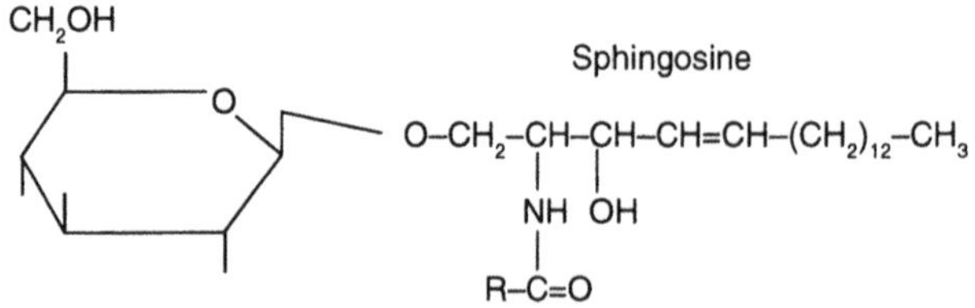

FIGURE 8.1: *1-β-glucosyl ceramide, one of the two cerebrosides (the other being 1-β-galactosyl ceramide), which are the simplest glycosphingolipids. R–C=O is a fatty acyl residue usually with a straight chain of over 20 carbon atoms, sometimes containing a hydroxyl group or a double bond. In a few tissues the sphingosine moiety is replaced by derivatives such as dihydrosphingosine or 4-hydroxy sphingosine (see Section 4.2.2).*

TABLE 8.1: *Monosaccharides commonly found in glycosphingolipids*

Common name	Abbreviation	Structure
Glucose	Glc	
Galactose	Gal	
N-Acetylglucosamine	GlcNAc	
N-Acetylgalactosamine	GalNAc	
Fucose	Fuc	
N-Acetylneuraminic acid (sialic acid)	Neu5Ac	

impenetrable. However, the semi-systematic nomenclature recommended by IUPAC/IUB is now largely accepted and is used here. It is based on a short list of simple, neutral glycosphingolipids (*Table 8.2*) which are assigned trivial names. Each more complex glycosphingolipid is a derivative of one member of this list and is named systematically, accordingly. The distance, measured in sugar residues, of each monosaccharide from the ceramide residue is indicated by a Roman numeral and the position on that monosaccharide of a substituent is indicated by a superscript Arabic numeral. In this way the ganglioside with the structure:

Galβ1-3GalNAcβ1-4Gal(3-2αNeuAc)β1-4Glcβ1-Cer

would be described unambiguously as:

II[3]N-acetylneuraminosyl gangliotetraosyl ceramide

or

II[3]Neu5Ac-Ggose$_4$Cer.

TABLE 8.2: *Basis of the IUPAC/IUB nomenclature of glycosphingolipids: names and abbreviations of simple glycolipids*

Structure	Trivial name of oligosaccharide[a]	Symbol[b]	Short symbol[c]
Gal(α1-4)Gal(β1-4)GlcCer	Globotriaose	GBOse3	Gb3
GalNAc(β1-3)Gal(α1-4)Gal(β1-4)GlcCer	Globotetraose	GbOse4	Gb4
Gal(α1-3)Gal(β1-4)GlcCer	Isoglobotriaose	iGbOse3	iGb3
GalNAc(β1-3)Gal(α1-3)Gal(β1-4)GlcCer	Isoglobotetraose	iGbOse4	iGb4
Gal(β1-4)Gal(β1-4)GlcCer	Mucotriaose	McOse3	Mc3
Gal(β1-3)Gal(β1-4)Gal(β1-4)GlcCer	Mucotetraose	McOse4	Mc4
GlcNAc(β1-3)Gal(β1-4)GlcCer	Lactotriaose	LcOse3	Lc3
Gal(β1-3)GlcNAc(β1-3)Gal(β1-4)GlcCer	Lactotetraose	LcOse4	Lc4
Gal(β1-4)GlcNAc(β1-3)Gal(β1-4)GlcCer	Neolactotetraose	nLcOse4	nLc
GalNAc(β1-4)Gal(β1-4)GlcCer	Gangliotriaose	GgOse3	Gg3
Gal(β1-3)GalNAc(β1-4)Gal(β1-4)GlcCer	Gangliotetraose	GgOse4	Gg4
Gal(α1-4)GalCer	Galabiose	GaOse2	Ga2
Gal(1-4)Gal(α1-4)GalCer	Galatriaose	GaOse3	Ga3
GalNAc(1-3)Gal(1-4)Gal(α1-4)GalCer	N-Acetylgalactos-aminylgalatriaose	GalNAc1-3GaOse$_3$	—

[a] Name of glycolipid is formed by converting ending '-ose' to '-osyl', followed by '-ceramide' without space; e.g. globotriaosylceramide.
[b] Should be followed by Cer for the glycolipid, without space; e.g. McOse3Cer, McOse4Cer.
[c] The short form should be used only in situations of limited space or in case of frequent repetition.

In fact each of these glycolipids will probably be a small family of closely related structures differing in the nature of the fatty acyl residue on the sphingosine. Heterogeneity of this residue involves variability in chain length, branching and degree of unsaturation. In some cases the sphingosine moiety may also be modified slightly (see Chapter 4, Section 4.2.2). The analysis of these complex compounds has been discussed recently by Schnaar [2].

8.1.3 Detection

Glycosphingolipids do not possess chromophoric groups that allow detection by color. When present in UV-transparent solvents they may be detected nonspecifically by instruments capable of detecting UV light in the 210–215 nm region. This is a feature of several HPLC systems. Alternatively, the presence of monosaccharide units allows detection by one of several color reactions. Most of these depend on hydrolysis and dehydration to furfural and its derivatives by heating with strong acids. These products give characteristically colored condensation products with a wide range of reagents. On the other hand, the treatment of aminosugars, including sialic acids, with alkali provides a basis for their specific detection. Oxidation of sugars with unsubstituted hydroxyl groups on adjacent carbon atoms (i.e. vicinal hydroxyl groups) with periodate has also given rise to useful methods of detection, especially of sialic acids (free or conjugated).

Detection methods based on strong acid treatment have utilized a wide range of condensation reagents for color development. Anthrone gives a color with most types of sugars and is often used as a general carbohydrate reagent. Combined with strong sulfuric acid a green color is produced both in solution or on the surface of a thin layer chromatogram.

Several phenolic reagents have been used in the detection of furfural derivatives produced from sugars. This is the most commonly used group of chromogenic substances in this area and includes phenol, naphthol, orcinol, resorinol and naphthoresorcinol. For example, Bial's reagent employs orcinol and Fe^{3+} in concentrated aqueous hydrochloric acid to detect microgram quantities of sialic acid by producing a blue-purple color soluble in amyl alcohol. A modified Bial's reagent using resorcinol and Cu^{2+} is reputed to be slightly more sensitive. Naphthol in strong acid gives deep red-purple colored spots with 1–10 µg of glycosphingolipids on thin layer chromatograms.

Color tests for sugars involving alkaline treatment depend on the production of pyrroles and other related compounds that generate a pink to purple color when further treated with an acidified solution of p-dimethylaminobenzaldehyde (Erhlich's reagent). The Elson–Morgan test for hexosamines (free or combined) is based on their reaction in alkaline solution with ethyl acetoacetate or 2,4-pentanedione to form pyrroles which give a pink color with Ehrlich's reagent. The Morgan–Elson test for free hexosamines is simpler, in that after heating with alkali direct mixing with Ehrlich's reagent gives an intense purple color. Methods based on the Ehrlich reaction have been developed which have high specificity for both free sialic acid and for sialosyl compounds. They are, however, less sensitive than those employing Bial's reagent.

Periodate oxidation of sugars containing vicinal hydroxyl groups (i.e. hydroxyl groups on adjacent carbon atoms) results in the production of aldehydes that can be detected by several reagents. One of the best known tests based on this method is the periodic acid-Schiff (PAS) reaction. Schiff's reagent (acidic, reduced fuchsin) condenses with aldehydes to give a pink product. It is not very sensitive to neutral glycolipids but is more sensitive to polysialo-compounds. Malonaldehyde and 3-formylpyruvate are major products of periodate oxidation of both sialic acids and of 2-deoxypentoses. Both react with thiobarbituric acid to give intense red colors. Providing samples are free of 2-deoxysugars and lipids that give rise to malonaldehyde on oxidation with periodate, this can provide a very sensitive test for sialic acids.

All of these color tests can be applied to solutions or suspensions and also as stains to thin layer chromatograms to demonstrate the presence of sugars. The source, nature and level of purity of the sample will be important in deciding if the sugar forms part of a glyco-sphingolipid. Some of the color tests (e.g. Morgan–Elson test) require the monosaccharide to be detected to be released from the glycolipid before they will work.

Some proteins such as lectins and antibodies may bind specifically and strongly to the carbohydrate portion of glycosphingolipids and have proved useful in detecting these compounds on thin layer chromatograms and, occasionally, on cell surfaces. For example, the lectin of *Mackia amurensis* is specific for α-2,3-linked sialic acid, those of *Arachis hypogea* and *Allomyrina dichotoma* are both specific for terminal β-galactose, and that of *Erythrina crystagalli* is specific for terminal α/β-galactose and α/β-*N*-acetylgalactosamine. These, and other compounds, are available covalently linked to fluorescent labels, which can be detected directly, or to enzymes such as peroxidase or phosphatase, which can be detected at high sensitivity by converting a colorless substrate to a colored product. Amplification of the response and therefore sensitivity can be increased by using a poly-clonal antibody specific to several epitopes of the lectin, the antibody being tagged with a fluorescent material or a detectable enzyme. The use of lectins linked to biotin is also popular. The biotin is then detected by specific and strong binding with avidin which is itself labeled with a fluorescent tag or with an enzyme. The use of antisera to detect specific antigenic glycosphingolipids, e.g. some blood group determinants, has been described. The method is very similar to that using lectins employing analogous visualization and amplification procedures. Several ingenious modifications of this general theme exist that are proving sensitive and specific for detecting glycosphin-golipids.

8.1.4 Isolation and purification

Initial extraction of glycosphingolipids from tissues is performed according to the method of Folch (see Chapter 2). Attempts to maximize extraction of anionic and polar derivatives then usually involves re-extraction using more polar solvent mixtures: for example, chloroform–methanol 1:1 (v/v), followed by 1:2 (v/v). Some workers prefer to add aqueous, 0.4 M sodium acetate to the latter mixture (1:12 by vol.) to facilitate extraction of anionic lipids. Other successful procedures have involved phosphate-buffered solvents such as butanol, propanol–hexane–water mixtures or tetrahydrofuran–water mixtures.

The method of Svennerholm devised for gangliosides is presented in the flow sheet, *Figure 8.2*. If large quantities of tissue (kilograms) are to be extracted it may be advantageous to remove first much of the unwanted lipids by acetone extraction. A suspension of the resulting acetone-dry powder in water (1:1 w/v) may then be treated as if it were homogenate in *Figure 8.2*, or more simply the powder is extracted with hot 95% ethanol for 10–30 min.

Partitioning of the extracts continues as in the lower part of the flow sheet. A similar solvent partition is also usually needed if a Folch extraction was initially carried out. A very effective parititioning system for gangliosides has been described in which the initial total lipid extract is dried solvent-free and then redissolved in a mixture of di-isopropyl ether–butanol (3:2 v/v, 20 vol. per gram of tissue). This is partitioned with aqueous sodium chloride (50 mM, 10 vol.). The resulting lower phase is re-extracted with the organic solvent (20 vol.) to leave most of the gangliosides in the lower phase.

Whichever of these methods is adopted, it is helpful to monitor the process by a simple color test, say for sialic acids, or by TLC. The amount of other lipids, acids and salts in the tissue may well influence the efficacy of the process and necessitate some modification. Before further processing, e.g. chromatography of the components of the aqueous layer, it will be essential to desalt the extract, either by dialysis or by size exclusion gel chromatography.

Treatment of a lipid extract with mild alkali will hydrolyze *O*-acyl but not *N*-acyl linkages. For example, the fatty acyl groups of phospholipids, triacylglycerols and cholesterol esters will be liberated as salts by this procedure but the fatty acyl group of sphingosine will not. This will facilitate removal of these lipid contaminants from the glycosphingolipids by converting them to products that are water soluble and dialyzable or separable more readily by silicic acid chromatography. However, it will also remove *O*-acetyl groups from sialic acids and should not be used if analysis of *O*-acetyl patterns of these sugars is anticipated.

Bulk separations of glycosphingolipids by column chromatography often employ silicic acid and gradient elution by increasing stepwise the concentration of methanol in chloroform from 2 to 67% (v/v) – the latter mixture eluting the glycosphingolipids. Loadings of 1 g of lipid per 10 g of silicic acid are quite common. Smaller quantities of lipid

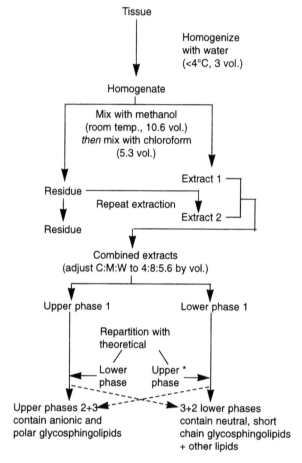

FIGURE 8.2: Flow sheet summarizing a procedure for the isolation and purification of glycosphingolipids. * The presence of 10 mM aqueous KCl instead of water assists separation.

may be handled conveniently using small cartridges such as Sep-Paks® of silicic acid and similar stepwise gradients of solvent.

Further fractionation of the glycosphingolipid fraction may then be achieved by employing one of several options. If silicic acid is still the preferred chromatographic material, finer separations may result from classical column chromatography, using a small loading on the column and a more refined solvent gradient. Gradients such as chloroform–methanol–water, 53:45:2 (by vol.) to 25:73:2, or 2-propanol–hexane–water, 50:40:5 (by vol.) to 55:30:15 have been described. Monitoring of the eluent by sugar color tests or by TLC is recommended. Alternatively, preparative TLC on silica gel has also been successful using similar solvent mixtures. Rapid highly sensitive detection can be achieved by spraying with dilute primuline and viewing under UV light. The light bands of lipids, against a dark background, can then be scraped from the plate and extracted with chloroform–methanol–water, 10:10:1 (by vol.) or with 2-propan-ol–hexane–water, 55:25:20 (by vol.). Inevitably, some contaminants from the silica gel and primuline will be present in the final samples and may be best removed by small reversed phase Sep-Pak columns. Preparative HPLC on silica can be very successful especially if 2-propanol–hexane–water mixtures are used which allow the separation to be monitored at wavelengths close to 210 nm.

Preparative resolution of anionic glycosphingolipids can also be achieved by anion exchange chromatography on DEAE Sepharose or on Q-Sepharose. The resin is usually first converted to the acetate form. Neutral lipids are then eluted with, say, chloro-form–methanol–water, 30:60:8 (by vol.) followed by anionic glyco-lipids, when the water in the eluent is replaced by 0.8 M aqueous sodium acetate. This bulk elution of glycosphingolipids may be refined by a more gradual gradient, resulting in separation of several individual components with respect to carbohydrate structure. If the co-chromatography of glycosphingolipids with contaminating lipids such as phospholipids proves difficult it may well be worth combining peracetylation with one of the chromatographic procedures. In this approach the polarity of the glycolipids is much reduced by acetylation of all of the hydroxyl groups of the sugar moiety using a mixture of pyridine and acetic anhydride. This changes the chromatographic mobility of the glycosphingolipids so that they are easily separated from the other contaminating lipids. For example, glycosphingolipids which are eluted from a silicic acid column only by a 1:3 (v/v) mixture of chloroform–methanol can be eluted by a 19:1 (v/v) mixture after peracetylation. After elution from the column the peracetyl-glycosphingolipids are deacetylated by mild treatment with KOH in a toluene–methanol mixture. This procedure cannot be used if natural O-acetyl sugars in glycolipids are to be investigated, as the final treatment deacetylates these also. Fatty acyl groups on the lipid are also lost by this procedure.

Confirmation of the nature of the glycan moiety of a glycosphingolipid may be greatly assisted by the use of specific hydrolytic enzymes. Several exoglycosidases (including α-fucosidase, α- and β-N-acetyl-hexosaminidase and α-neuraminidase) are now available which are specific to both the nature of the terminal monosaccharide and the anomeric configuration of its glycosidic linkage. Few of these are specific to a particular linkage, but two different neuraminidases showing preference for $\alpha(2 \rightarrow 3)$ or $\alpha(2 \rightarrow 6)$ are available commercially. Some glycosphingolipids are not substrates for these exoglycosidases. In this situation it may be helpful to use ceramide glycanase which releases the entire glycan moiety if it is linked to ceramide via glucose (as in most glycosphingolipids). The released glycan may then be more amenable to exoglycosidase attack and to other glycoprotein glycan methods.

The resolution of microheterogeneity of glyosphingolipids due to variations in the fatty acyl chain of the ceramide moiety is probably best attempted using reversed phase partition chromatography. For example, successful preparative HPLC of <0.5 mg of ganglioside on a C_{18} analytical HPLC column using mixtures of methanol and water or of acetonitrile and aqueous sodium phosphate, detecting the solute by absorbance of UV light at 195–205 nm, has been reported.

Because of the complexity of glycan structures in glycosphingolipids, co-chromatography with a particular standard compound of known structure in several different chromatographic systems does not always guarantee the identity of the structure. Ideally structural confirmation requires the rigor and sensitivity of mass spectrometry allied to other information such as nuclear magnetic resonance. Fortunately this ideal is now a feasible proposition, with several procedures involving on-line mass spectrometry on automated chromatographic systems being developed. Analysis of the linkages between sugar residues of a glycan moiety of a glycolipid can be achieved on picomolar amounts using a microscale methylation analysis in which the products are identified unambiguously and quantified by GC–MS. In this procedure all free hydroxyl groups of the glycan are converted to their methyl ethers by base-catalyzed permethylation. Lithium dimethylsulfinyl carbanion and methyl iodide is a favored reaction mixture. The permethylated glycan is then subjected to acid hydrolysis followed by reduction of the partially methylated monosaccharides with sodium borohydride. Peracetylation follows using acetic anhydride and pyridine to give a mixture of methylated alditol acetates which are separated and identified by GC–MS. As shown in *Figure 8.3* the position of acetate residues indicates the position of each glycosidic linkage in the original glycan.

Fast atom bombardment mass spectrometry (FABMS) is particularly useful in providing informative fragmentation patterns. Generally, in order to obtain fragmentation patterns that can be assigned unambiguously, it is necessary to make derivatives. Usually functional groups are protected by permethylation or peracetylation. Sometimes other tags are also added to increase sensitivity. Accurate calculation of the molecular masses and fragment ions provides evidence of both composition and sequence. Occasionally linkage information may also emerge from the spectra, but GC–MS of methylation analysis products is more likely to provide this. Mass spectrometry is clearly a very powerful tool if structural information is needed, especially when combined with other more classical chemical approaches. However, it is expensive in terms of hardware and demands expert operation and interpretation. The average laboratory may well not have the necessary facilities and expertise to take the studies this far.

8.1.5 Quantitation

A sphingosine assay will give the molar amount of glycosphingolipid present. A popular, sensitive method depends upon the liberation of the sphingosine by acid methanolysis of the glycosphingolipid followed by treatment of its solution in ether with fluorescamine and by fluorescent spectrometry (see also Chapter 4, Section 4.2.5).

Several other accurate assay methods for glycosphingolipids are developments of the techniques used for their detection. For example, the various color tests for sugars can be made into quite sensitive, quantitative methods by measuring the intensity of color developed in solution under standardized conditions. In most animal cells glycosphingolipids are present in much higher concentration than other glycolipids. It follows that assay of lipid-bound sugars will reflect primarily the glycosphingolipids present. Anionic glycosphingolipids are probably best assayed by measuring lipid-bound sialic acid. This can be measured with high sensitivity by Bial's test, Ehrlich's test or by the periodate–thiobarbituric acid method. Knowledge of the type of anionic glycosphingolipid present then enables an assessment of the molar quantity of the compound. Neutral glycosphingolipids are usually assayed as lipid-soluble neutral sugar by a general assay based on the strong acid–phenolic reagent tests. In all of these assays it is important to indicate the nature of the standard sugar(s) used for calibration.

Information on the amount of individual compounds clearly requires a combination of chromatographic and assay methods. Possibly the most sensitive combination is HPLC, several systems of which lead to good resolution of mixtures into their component compounds, which can then be assessed accurately by on-line UV absorption measurements at 205–215 nm. Systems based on amino-bonded silica and underivatized silica have been described. UV-transparent solvents are required and combinations of aqueous phosphate buffer and acetonitrile or 2-propanol and water have proved satisfactory, especially for gangliosides in subnanomolar amounts.

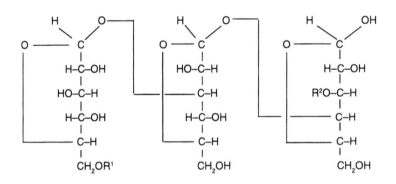

R¹6Glc1-3Man1-4(R²1-3)Gal

↓ Methylation

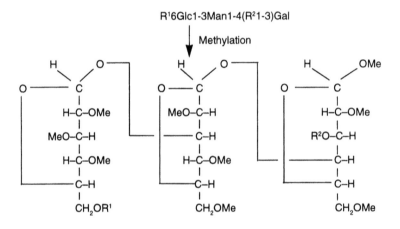

↓ Hydrolysis

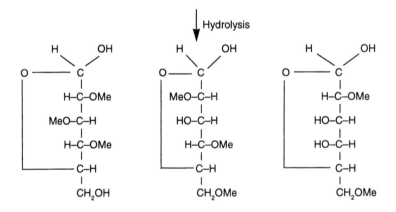

↓ Reduction

(continued on following page)

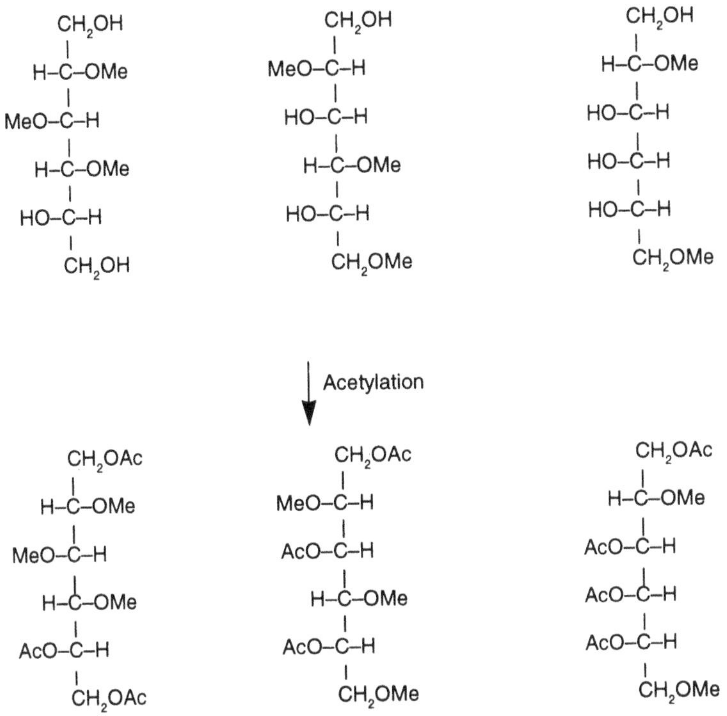

CH$_2$OH
|
H–C–OMe
|
MeO–C–H
|
H–C–OMe
|
HO–C–H
|
CH$_2$OH

CH$_2$OH
|
MeO–C–H
|
HO–C–H
|
H–C–OMe
|
HO–C–H
|
CH$_2$OMe

CH$_2$OH
|
H–C–OMe
|
HO–C–H
|
HO–C–H
|
HO–C–H
|
CH$_2$OMe

↓ Acetylation

CH$_2$OAc
|
H–C–OMe
|
MeO–C–H
|
H–C–OMe
|
AcO–C–H
|
CH$_2$OAc

CH$_2$OAc
|
MeO–C–H
|
AcO–C–H
|
H–C–OMe
|
AcO–C–H
|
CH$_2$OMe

CH$_2$OAc
|
H–C–OMe
|
AcO–C–H
|
AcO–C–H
|
AcO–C–H
|
CH$_2$OMe

1,5,6 triacetyl
2,3,4 trimethyl
glucitol

1,3,5 triacetyl
2,4,6 trimethyl
mannitol

1,3,4,5 triacetyl
2,6 dimethyl
galactitol

FIGURE 8.3: *Illustration of the use of methylation of glycans in the analysis of glycosidic linkages. Assessment of the acetylation and methylation pattern by GC–MS of the glucitol, mannitol and galactitol derivatives produced allows deduction of the types of linkages in the trisaccharide shown.*

Glycosphingolipids are not sufficiently volatile to allow GLC assays. However, it is possible to release the monosaccharides by trifluoroacetolysis, and these can then be analyzed by GLC of their more volatile alditol acetates or trimethylsilyl derivatives. Alternatively, the released monosaccharides may be measured directly with comparable sensitivity by high pH anion exchange chromatography (HPAEC) using Dionex equipment fitted with a pulsed amperometric detector. At a pH of 12 all hydroxyl groups on sugar molecules ionize giving a charge distribution that differs slightly from one monosaccharide to another. This ionization is the basis of both the high resolution of the chromatographic technique and of the high sensitivity of the detector.

8.2 Polyisoprenyl phosphosugars

8.2.1 Biological significance

It was mentioned in Chapter 4 that polyprenols and dolichols are involved in some glycosyl transfer reactions. Their phosphate derivatives function as membrane-bound carriers of mono- and oligosaccharides in the assembly of complex glycans and in their translocation across the membrane. In bacteria, bactoprenyl (undecaprenyl) phosphate acts as an essential co-factor in the bacterial membrane located synthesis of several cell wall polysaccharides (*Figure 8.4*). In eukaryotic organisms, dolichyl phosphate functions analogously as a coenzyme in the endoplasmic reticulum-mediated *N*-glycosylation of proteins (*Figure 8.5*). In yeast and fungi it is also important in *O*-mannosylation of proteins. Polyisoprenyl phosphosugars and dolichyl phosphosugars are formed as intermediates in these glycosylation processes and hence rarely accumulate. Quite often their detection and assay requires the use of radioisotopically labeled precursors or the aid of mutant cells lacking a particular (late) step in the biosynthetic schemes shown in *Figures 8.4* and *8.5*. The use of cell-free membrane preparations may also slow down the overall process, allowing measurable quantities of the products of individual steps to accumulate.

8.2.2 Structures

Examples of the sorts of structures found as intermediates in the biosynthesis of bacterial wall polymers and in eukaryotic *N*-glycosylation of proteins are given in *Figures 8.6* and *8.7* respectively.

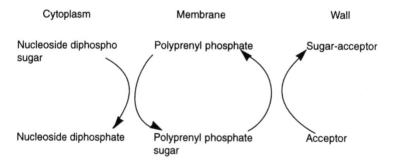

FIGURE 8.4:   *The generalized role of polyprenyl phosphate sugars in bacterial cell wall synthesis.*

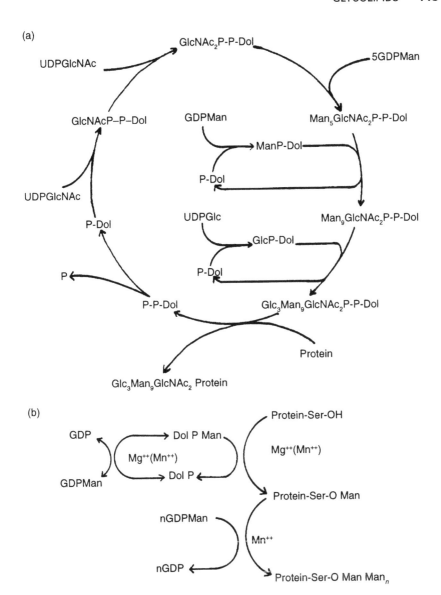

FIGURE 8.5: *The role of dolichyl phosphate sugars in (a) N-glycosylation of proteins and (b) O-mannosylation of yeast and fungal proteins. Dol, dolichol; P, phosphate; UDPGlcNAc, uridine diphospho-N-acetylglucosamine; GDPMan, guanosine diphosphomannose; UDPGlc, uridine diphosphoglucose. See* Table 8.1 *for other abbreviations.*

8.2.3 Detection

Usually the concentration of these compounds in bacterial or eukaryotic cells is very low and their detection requires the use of

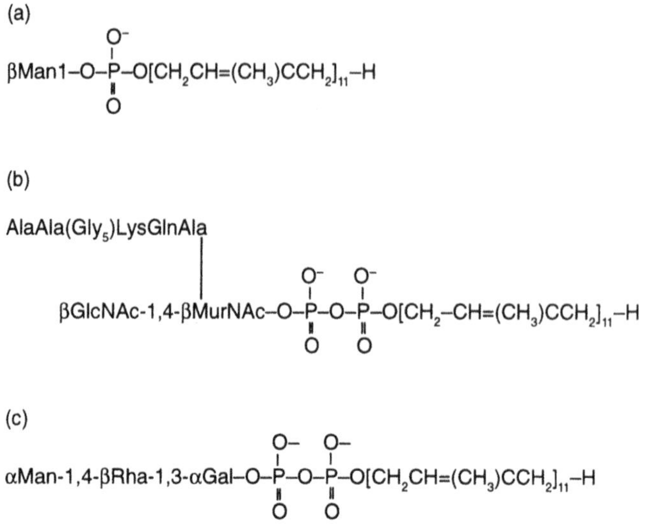

(a)

$$\beta Man1-O-\overset{\overset{O^-}{|}}{\underset{\underset{O}{\|}}{P}}-O[CH_2CH=(CH_3)CCH_2]_{11}-H$$

(b)

AlaAla(Gly$_5$)LysGlnAla

$$\beta GlcNAc\text{-}1,4\text{-}\beta MurNAc-O-\overset{\overset{O^-}{|}}{\underset{\underset{O}{\|}}{P}}-O-\overset{\overset{O^-}{|}}{\underset{\underset{O}{\|}}{P}}-O[CH_2-CH=(CH_3)CCH_2]_{11}-H$$

(c)

$$\alpha Man\text{-}1,4\text{-}\beta Rha\text{-}1,3\text{-}\alpha Gal-O-\overset{\overset{O^-}{|}}{\underset{\underset{O}{\|}}{P}}-O-\overset{\overset{O^-}{|}}{\underset{\underset{O}{\|}}{P}}-O[CH_2CH=(CH_3)CCH_2]_{11}-H$$

FIGURE 8.6: *Examples of bacterial polyisoprenyl phosphosugars: (a) undecaprenyl monophosphate mannose, an intermediate in cell wall mannan synthesis; (b) undecaprenyl diphosphate N-acetylglucosaminyl (pentapeptide) N-acetyl-muramic acid, an intermediate in cell wall peptidoglycan synthesis; (c) undecaprenyl diphosphate mannosyl rhamnosyl galactose, an intermediate in cell wall antigenic lipopolysaccharide synthesis. Abbreviations: MurNAc, N-acetylmuramic acid; Rha, rhamnose; Man, mannose; standard amino acids; all others as in Table 8.1.*

radioisotopically labeled precursors. The glycosylated moieties can be labeled selectively by incubating cells or tissue slices with a short pulse of radioactive monosaccharide of high specific radioactivity. Homogenates of the cells or tissues are then extracted with organic solvent, usually chloroform–methanol (2:1 v/v) for complexes of short chain glycans with undecaprenyl or dolichyl phosphates. For complexes with larger chain glycans (e.g. *Figure 8.7*) a more polar solvent such as chloroform–methanol–water (1:1:0.3 by vol.) will probably be necessary.

These solubility characteristics are usually retained after treatment with mild alkali, but after exposure to mild acid treatment (pH 1, 10 min, 100°C) the radioactivity is rendered water-soluble again. This is because of the acid lability of the sugar-1-phosphate (*Figures 8.6* and *8.7*) and of the allylic phosphate (*Figure 8.6*) bond linking the water-soluble glycan to the lipid-soluble polyprenyl chain. The effect of mild acid/alkali treatment on their solubility in chloroform–methanol mixtures is usually sufficient to detect the presence of glycan

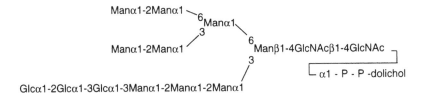

FIGURE 8.7: *The donor oligosaccharyl diphosphate dolichol that is involved in the final step of protein* N-*glycosylation (Figure 8.5) (see* Table 8.1 *for abbreviations).*

phosphoprenols (dolichols). Further identification may involve TLC against standard compounds on silica using chloroform–methanol mixtures. If the experiment requires identification primarily of the (radioactive) glycan moiety, chromatography on a sizing column (e.g. Bio-Gel P-4) of the water-soluble, mild acid hydrolysis products can be employed. This system is capable, for example, of resolving the glycans from most of the intermediates in the dolichyl phosphate pathway of N-glycosylation (*Figure 8.5a*). Further details of analysis of this complex group of compounds is beyond the scope of this book.

8.2.4 Isolation and purification

The methods described by Behrens and Tabora in 1978 [3] still form the basis of the chromatographic separation of most glycosylated derivatives of polyprenyl and dolichyl phosphates. Essentially this involves chromatography on DEAE cellulose acetate using a gradient of ammonium formate (or acetate) in chloroform–methanol–water (1:1:0.3). Neutral lipids are eluted by the solvent alone. Monophospho-derivatives are eluted earlier in the gradient (usually around 0.02 M) than are diphospho-derivatives (usually around 0.06 M). Preparative TLC (especially HPTLC) may then successfully eliminate most of the impurities.

8.2.5 Quantitation

Measurement of the amount of polyisoprenyl phosphosugars formed by micro-organisms or animal or plant cells in culture usually involves incubation with a radioactive sugar or phosphate as precursor in the medium until equilibration with endogenous compounds has occurred. The product of interest may then be extracted and subjected to TLC against markers, to check the chromatographic homogeneity of the product, and some purification, as indicated in Section 8.2.4, may be necessary. The amount of radioisotope recovered in the radioisotopically and chromatographically homogeneous product is recorded. If the radioisotope is associated with a sugar, then the amount of chloroform–methanol (CM)-soluble radioactivity which is sensitive to mild acid treatment should be determined. If radioactive phosphate is used as the precursor, then, in the case of unsaturated polyprenyl phosphosugars, the loss of radioactivity from the CM-soluble portion following treatment with mild acid should be determined. In the case of dolichyl mono[^{32}P] phosphosugars, the ^{32}P will remain with the dolichol (and hence be recovered in the CM-soluble fraction) but will have a different chromatographic mobility. Dolichyl di[^{32}P]-phosphosugars will, upon mild acid treatment, retain half of the ^{32}P as dolichyl monophosphate.

8.3 Glycosylglycerides and related compounds

8.3.1 Biological significance

Glycosylglycerides are major lipid components of the photosynthetic membranes of higher plants and of algae [4]. Chloroplast membranes of higher plants contain about 40% (by dry weight) of a mixture of mono- and digalactosyldiacylglycerols. This is usually accompanied by smaller quantities of sulfoquinovosyldiacylglycerol (see also sulfolipids, Chapter 7). Photosynthetic membranes of algae, especially of marine species, yield high concentrations of similar compounds containing glucose as the main sugar. Although the high concentrations of these glycosylglycerides must greatly influence the properties of these biologically important membranes, their specific functions are not understood.

Diglycosyldiacylglycerols also occur widely but at much lower concentrations in the membranes of Gram-positive and photosynthetic Gram-negative bacteria. They have not been reported in animals. However, traces of monogalactosyldiacylglycerol have been isolated from brain and testes where they have been accompanied by the corresponding monoether analog monogalactosylalkylacylglycerol. Some bacteria contain small amounts of glycosyldialkylglycerols. Again, the function of these compounds is uncertain.

8.3.2 Structures

The sugar moiety of the glycosylglycerides is linked glycosidically to position 3 of glycerol with the other positions carrying fatty acyl groups (*Figure 8.8*). Variations in the sugar portion of those compounds derived from plants include monogalactosyl, digalactosyl and sulfoquinovosyl (*Figures 8.8a, b* and *c* respectively). Glucosyl derivatives (for example *Figure 8.8d*) are found in algae. Several mono- and diglycosylglycerides are found in bacteria (*Figures 8d–h*); the diglycosyl versions usually predominate.

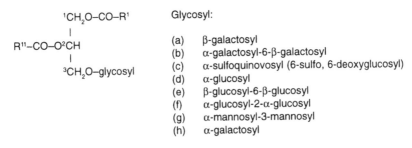

FIGURE 8.8:   *Structure of some major glycosyldiglycerides of plants, algae and bacteria carrying carbohydrates (a)–(h) as the glycosyl residue where R'CO and R''CO are fatty acyl residues.*

In plants the fatty acyl groups often consist mainly of γ-linolenic acid. Any palmitic acid or other saturated fatty acid present is usually esterified at position 2 of the glycerol moiety. The ether derivatives 3-glycosyl, 1-alkyl, 2-acyl glycerol and 3-glycosyl 1,2-dialkyl glycerol have been isolated principally from bacteria but only in very small amounts.

Although not strictly members of this group of glycolipids, it is convenient to mention here also acylated sugar derivatives, glyco-sylated hydroxy fatty acids and sterol glycosides (see also Section 4.1.2 which may accompany glycosylglycerides in lipid extracts. For example, 3,4,6-trioleoyl glucose and α-rhamnosyl-2-α-rhamnosyl-3-hydroxydecanoic acid have been described in bacterial lipid extracts. Many plants contain sterol glycosides in which a glucose residue is linked glycosidically to the 3-hydroxyl group of the sterol. In some cases the glucose may be replaced by another hexose. A fatty acyl residue may also be esterified to position 6 of the hexose.

8.3.3 Detection

Glycosylglycerides possess no chromophoric groups but a relatively strong absorption of UV light in the 200–215 nm region results from the unsaturated nature of the fatty acyl chains and allows their detection in UV-transparent solvents. In this and other respects methods of detection are very similar to those used for the glycosphingolipids discussed in Section 8.1.3. Sterol glycosides can also be detected by the Liebermann–Burchard reaction (see Section 4.3).

8.3.4 Isolation and purification

Although most members of this group of compounds can be extracted by the method of Folch (see Chapter 2) several workers prefer to ensure complete recovery by boiling homogenates of tissue/cells with a polar solvent mixture such as propanol–chloroform–methanol (1:2:1 by vol.). The extract may then be partitioned with an aqueous solution of sodium chloride (0.45% w/v) and the organic phase subjected to preparative TLC on silica gel. Solvents such as chloroform–methanol (9:1 or 7:3 by vol.) have been used successfully, the more polar preparations being appropriate for the diglycosyldiglycerides. Some workers prefer to have acetic acid present especially if anionic lipids are involved. In this context the mixture chloroform–methanol–acetic acid–water (85:15:10:3.5 by vol.) has given good separations of sulfoquinovosyldiglycerides from other lipids.

If large quantities of extracted lipid are to be handled it may be preferable to make the preliminary separation of lipid families by adsorption chromatography on a column of silicic acid. After elution of nonpolar lipids by chloroform a mixture of chloroform–acetone (1:1 by vol.) elutes the monogalactosyldiglycerides, whilst acetone alone elutes the digalactosyldiglyceride and some sulfoquinovosyldiglyceride. The remaining part of the sulfoquinovosyldiglyceride is eluted by chloroform–methanol (1:1 by vol.) [4].

Usually this preliminary separation of glycosylglycerides and related compounds into families of compounds is followed by reversed phase HPLC, for example, on Spherisorb C_6 using a gradient of 50–100% acetonitrile in water. In this system each family of compounds (e.g. monogalactosyldiglyceride or digalactosyldiglyceride) gives rise to several peaks with characteristic retention times dependent upon the mixture of fatty acyl residues of the glyceride moiety. The satisfactory separation of sulfoquinovosyldiglycerides requires the prior methylation of the acidic group by diazomethane. In this reversed phase HPLC system elution is usually monitored at 200–210 nm. However, dipalmityl and other saturated diacyl glyceride derivatives escape detection at this wavelength. Their presence in a family of compounds can be demonstrated after the TLC stage by replacing the glycosyl residue with a *p*-anisoyl residue. The resulting 1,2-diacyl-3-*p*-anisoyl-glycerols can then be separated by reversed phase HPLC as above, but this time monitoring at 250 nm. The individual anisoylglycerides can be recovered separately. In studies on saturated glycosyldiglycerides it is essential that phospholipids are removed at the TLC stage for they also will form anisoylglycerides in this method. GLC of intact glycosylglycerides as their trimethylsilyl ether derivatives can be successful but better information may be obtained from GLC of the 3-acetyl derivative of the diglyceride released by acid hydrolysis.

Preparative TLC followed by reversed phase HPLC as described above is also appropriate for the purification of sterol glycosides.

8.3.5 Quantitation

As for the glycosphingolipids, the various color tests for detection can be developed into quite sensitive and reliable assays for glyco-sylglycerides separated by appropriate chromatographic procedures. However, the easiest and most reliable assay systems are probably those based on reversed phase HPLC as described in Section 8.3.4.

Liberation of fatty acids as their methyl esters for individual quantitation is best achieved by treatment with sodium methoxide, methanolic hydrogen chloride or boron trifluoride in methanol (10%, w/v) as mentioned in Chapter 6. Analytical GLC can then be used for sensitive identification and assay (see Chapter 5). The analysis of the carbohydrate portion of these molecules is probably best handled as described in Section 8.1.5 for glycosphingolipids.

8.4 Phosphoglycolipids, lipoteichoic acids and related compounds

8.4.1 Biological significance

A group of compounds exist in some Gram-positive bacteria that contain one or more monosaccharide residues linked through phosphate to diacylglycerol (or glycerol) [5]. These phosphoglycolipids are membrane bound, present in relatively small amounts and of unknown function. Lipoteichoic acids are found also in the membranes of some of this group of micro-organisms. These are closely related to the phosphoglycolipids and to the cell wall teichoic acids. Lipoteichoic acids are present in relatively high concentration and, by virtue of the polymeric nature of the teichoic acid portion, are poly-anionic. They may function as precursors of teichoic acids of the cell wall. In fact the long polymeric chain of glycerolphosphate renders them relatively insoluble in most organic solvents and, apart from describing their structures, the analysis of these compounds will not be considered.

Some other bacterial membrane lipids contain carbohydrate linked directly to glycerol as part of complex hydrophobic phosphate-containing compounds. For example, 3-phosphatidyl, 2-glucosaminyl glycerol has been found in some *Bacilli* and glycosyl, glycero-phosphoryl derivatives of dibiphytanyl diglycerol tetraether (in which two biphytanyl [C_{40}] chains join two glycerol residues) account for half of the total lipids of some methanogenic bacteria. Although of very limited distribution, this last group of compounds clearly must be considered to be important membrane constituents. The special biological feature that these glycolipids give to the membranes is not yet clear.

8.4.2 Structures

The phosphoglycolipids contain a monosaccharide residue which carries on carbon 6 a phosphatidyl group (or a glycerophosphate group) and which is linked glycosidically to position 3 of diacylglycerol, either directly or through another monosaccharide residue. For example, some *Streptococci* contain kojibiose (α-glucosyl-2, α-glucose) linking phosphatidic acid and/or glycero-1-phosphoric acid to diacyl glycerol as in *Figure 8.9a*.

The lipoteichoic acids (*Figure 8.9b*) consist of a polar teichoic acid portion made up of repeating units (30–40) of glycerol-1-phosphate linked through the terminal phosphate to the 6-position of a glucose residue. This is linked glycosidically to position 6 of a second glucose residue, which in turn is linked glycosidically to position 3 of diacylglycerol. The glycerophosphate portions may well be substituted with D-alanyl or glucosyl residues (*Figure 8.9b*).

8.4.3 Detection, isolation, preparation and quantitation

The analysis of these compounds is very similar to that of the glycosylglycerides discussed in Section 8.3. Good separations from

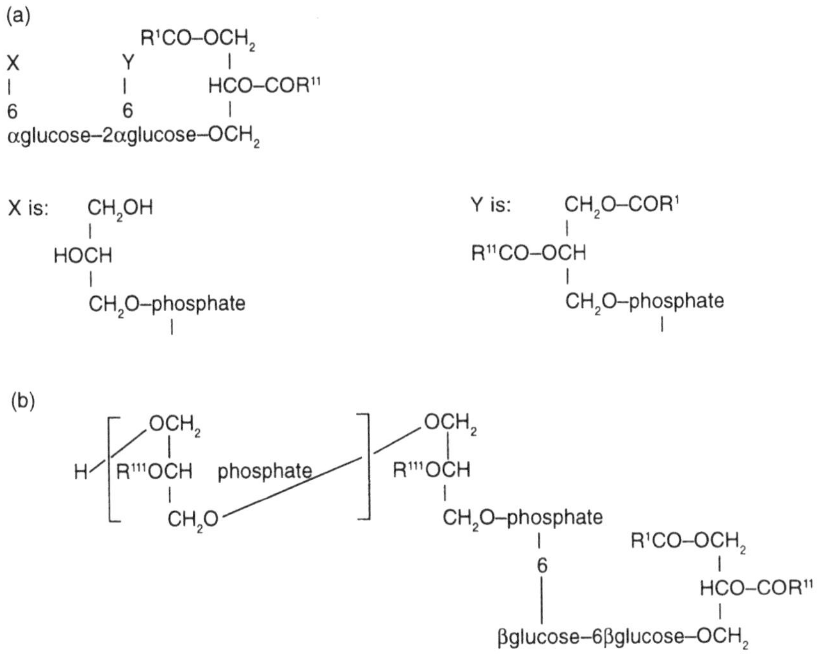

(a)

$$R^ICO-OCH_2$$

X   Y |
| | $HCO-COR^{II}$
6   6 |
αglucose–2αglucose–OCH_2

X is: CH_2OH Y is: CH_2O-COR^I
         | |
         HOCH $R^{II}CO-OCH$
         | |
         CH_2O–phosphate CH_2O–phosphate
            | |

(b)

$$\left[\begin{array}{c} OCH_2 \\ | \\ H \diagup R^{III}OCH \quad phosphate \\ | \\ CH_2O \end{array} \right] \begin{array}{c} OCH_2 \\ | \\ R^{III}OCH \\ | \\ CH_2O\text{–phosphate} \end{array}$$

            $R^ICO-OCH_2$
               |
      6       $HCO-COR^{II}$
               |
   βglucose–6βglucose–OCH_2

FIGURE 8.9:   Structures of (**a**) some phosphoglycolipids (substituents X and Y may both be present in the same molecule or present singly in separate molecules) and (**b**) lipoteichoic acids. R^ICO and $R^{II}CO$ are fatty acyl residues; R^{III} in (**b**) may be H, alanyl or glucosyl groups.

other impurities have been achieved by preparative TLC. The separated materials can then be analyzed for glucose, glycerol, phosphate and/or fatty acids as before.

References

1.   Schnaar, R.L. (1991) *Glycobiology,* **1,** 477–485.
2.   Schnaar, R.L. (1994) *Meth. Enzymol.,* **230,** 348–370.
3.   Behrens, N. and Tabora, E. (1978) *Meth. Enzymol.,* **50,** 403–435.
4.   Chapman, D.J. and Barber, J. (1987) *Meth. Enzymol.,* **148,** 294–319.
5.   Hancock, I.C. and Poxton, I.R. (1988) *Bacterial Cell Surface Techniques.* John Wiley & Sons, Chichester.

9 Lipoproteins

The transport of lipids in the blood from one tissue to another in animals has been facilitated by the evolution of several complexes of lipids and protein which are held together by noncovalent forces and are stable in an aqueous environment. These complexes are known as the plasma lipoproteins and they are the main subject matter of this chapter. They present special analytical features of their own but also bring together several aspects of individual lipid analysis dealt with primarily in Chapters 4–8.

Recently it has become increasingly clear that covalent lipid modification of many proteins is an important part of their post-translational modification with functional consequences. This includes the attachment of fatty acyl chains, prenylation and the addition of glycosylphosphatidylinositol groups. The analysis of fatty acyl proteins and of prenyl proteins is discussed here; glycosylphosphatidylinositol-linked proteins are discussed in Chapter 7.

9.1 Plasma lipoproteins

9.1.1 Structures and biological significance

Plasma lipoproteins [1] essentially exist as small spherical bodies with a surface layer of amphipathic phospholipids, glycolipids, cholesterol and glycoproteins. Combined within the sphere is a mixture of lipids, principally triacylglycerols and cholesterol esters. The amphipathic shell not only stabilizes the suspension of the hydrophobic contents in the aqueous blood plasma, but also carries signals in the glycoprotein components which enable different classes of lipoproteins to be recognized and bound specifically by particular cell types and by particular enzymes. These features are essential to the proper functioning of the lipoproteins in transporting and delivering lipids to tissues in response to their specific needs.

There have been several attempts to classify lipoproteins by virtue of their different physical and chemical properties and in a way that is helpful towards understanding their function. With regard to human plasma lipoproteins the most successful system has been based on the density of the particles as defined by their flotation characteristics upon density gradient ultracentrifugation. In humans there is a good correlation between the density of lipoproteins and their functions. Differences in density are brought about by differences in particle size and in chemical composition. These differences also enable separations to be achieved by electrophoresis, by gel chromatography and by polyanion precipitation. It should be emphasized that there is overlap between the different groups of lipoproteins, resulting in lipoprotein preparations that although predominantly of one 'type', are actually mixtures. Also, although there is considerable similarity of lipoprotein characteristics across mammals and other animal species, there are also important differences.

The main classes of lipoproteins are summarized in *Table 9.1* which also shows some average physical properties and chemical composition of each class prepared by density gradient centrifugation. The phospholipids present contain unsaturated fatty acids. It is

TABLE 9.1: *Some average physical and chemical properties of human plasma lipoproteins*

	Lipoproteins				
Property/component	CM	VLDL	IDL	LDL	HDL
Density range (g ml⁻¹)	0.90–0.95	0.95–1.006	1.006–1.019	1.019–1.063	1.063–1.210
Diameter (nm)	75–1200	30–80	25–35	18–25	5–12
Particle mass (Da × 10⁻⁶)	400	10–80	5–10	2.3	0.2–0.4
Phospholipids[a]	7	20	21	22	22
Cholesterol (+ ester)[a]	8	22	25	48	20
Protein[a]	2	7	10	20	50
Triacylglycerols[a]	83	50	40	10	8
Electrophoretic mobility[b]	Origin	Pre-β-	β-	β-	α-
Apolipoproteins (major types)	A,B₄₈	B,C,E (+C,E)	B,E	B₁₀₀	A,E (+C)

Lipoprotein acronyms are explained in the legend to *Figure 9.1* and in the text. Most of the phospholipids, cholesterol and proteins are to be found in the surface of the lipoprotein particle, while the majority of the cholesterol esters and triacylglycerols are present in an internal core. Derived primarily from ref. 1.
[a] % Particle mass.
[b] Related to mobility of globulins.

essential that they are protected from oxidative and free radical attacks. This is achieved largely by the presence of small quantities of the anti-oxidants ubiquinol and α-tocopherol. It can be seen in *Table 9.1* that the particle densities differentiate chylomicrons (CM) from very low density lipoproteins (VLDL), intermediate density lipoproteins (IDL), low density lipoproteins (LDL) and high density lipoproteins (HDL). The functions of these lipoproteins in the transport of triacylglycerols and cholesterol (esters) from one tissue to another and in their metabolism can be summarized as follows (*Figure 9.1*).

CM are synthesized in intestinal epithelial cells and are enriched in absorbed dietary triacylglycerols. They contain cholesteryl esters and may also contain fat soluble vitamins, all from the diet. When first formed CM carry only apolipoproteins of the A and B classes on the surface, but these are soon accompanied by apolipoproteins C and E derived from other lipoproteins (mainly HDL) in the circulation. Upon reaching the capillaries of adipose tissue (and some other organs such as muscle) CM bind to and activate a lipoprotein lipase, primarily through the agency of apolipoprotein C_2. This enzyme rapidly releases fatty acids (for tissue storage or oxidation) from the triacylglycerol. At the same time some of the surface components (some phospholipids and the apolipoproteins A and C) are transferred to HDL. This leaves particles called chylomicron remnants (still containing dietary cholesterol ester) which bind to an apolipoprotein E receptor on hepatocytes where they are internalized and degraded. This uptake is inhibited by apolipoproteins C present in unhydrolyzed CM, so restricting hepatic uptake to chylomicron remnants that have lost apolipoproteins C.

Triacylglycerols (and cholesterol esters) synthesized in the liver are also transported to adipose and other tissues. In this case the vehicle is VLDL which is synthesized in hepatocytes complete with apolipoproteins B, C and E. The apolipoproteins B of VLDL and LDL (B_{100}) are larger than those of CM (B_{48}) — the subscripts indicating molecular mass in kDa. Fatty acids are released in other tissues, especially adipose tissue, by processes analogous to those discussed for CM, i.e. by lipoprotein lipases on the surface of epithelial cells in capillaries activated by apolipoprotein C_2. Some phospholipids and apolipoproteins C are transferred to HDL. The remnant of this process (IDL) may be taken up by a hepatic IDL (apolipoprotein E) receptor or it may lose further triacylglycerol and phospholipids and apolipoproteins (except B_{100}) to produce a particle recognized as LDL. This particle is now particularly enriched in cholesterol esters originating from the liver and is taken up primarily in peripheral tissues through the LDL (apolipoproteins B/E) receptor as a result of possessing apolipoprotein B_{100}. Endocytosis and lysosomal release within peripheral cells of

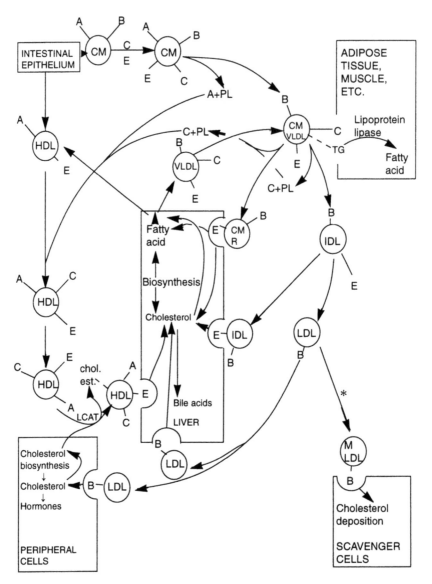

FIGURE 9.1: *Summary of the principle roles of plasma lipoproteins in the transport of triaclyglycerols (mainly the upper part of the figure) and cholesterol (esters) (mainly the lower part of the figure) in the blood circulation of humans. The lipoprotein particles are identified by their abbreviations inside small circles and the principal apolipoproteins (A, B, C or E) present are indicated. Abbreviations: CM, chylomicrons; VLDL, very low density lipoproteins; IDL, intermediate density lipoproteins; LDL, low density lipoproteins; HDL, high density lipoproteins; MLDL, modified low density lipoproteins; TG, triacylglycerol (of CM and VLDL); CMR, chylomicron remnants; LCAT, lecithin: cholesterol acyltransferase; chol. est., cholesteryl ester (of HDL); PL, phospholipid; *, modification (of LDL).*

increased LDL-cholesterol results in the switching off of cholesterol biosynthesis and of LDL receptor biosynthesis in those cells; these are essential features of cholesterol homeostasis.

Approximately two-thirds of LDL uptake is by this process. The rest appears not to involve receptors. Some of the LDL particles may be modified enzymically or by free radicals or superoxide before they lose most of their apolipoprotein E. These modified particles are not able to function properly as LDL and they are usually removed from the circulation through modified apolipoprotein B-sensitive (scavenger) receptors found, for example, on macrophages. Excessive diversion of LDL along this route is a feature of atherosclerosis.

Also important in cholesterol homeostasis are the HDL particles, which basically transport excess unesterified cholesterol from the plasma membranes of peripheral cells and from the surface of other lipoproteins to the liver. Nascent HDL contains apolipoproteins A and E, having been formed in both liver and intestinal epithelial cells. Further apolipoproteins A and C are obtained from CM and VLDL respectively. An essential part of this process is the enzyme lecithin : cholesterol acyltransferase (LCAT, *Figure 9.1*). LCAT is present in a subpopulation of HDL particles enriched in apolipoproteins A_1 and D. The lecithin (phosphatidylcholine) is present in HDL particles having been derived from IDL and CM as described above. The cholesterol having been esterified in this way may be exchanged with other lipoproteins and cell membranes through the activity of transfer proteins (e.g. cholesteryl ester transfer protein). Some of the HDL particles are taken up by the liver by a process involving an apolipo-protein E binding site on the apolipoprotein B/E (LDL) receptor. Lysosomal hydrolysis of the internalized cholesteryl esters releases unesterified cholesterol within hepatocytes for metabolism or esterification.

The complex nature of the transport of triacylglycerols and cholesterol (esters) in the blood (*Figure 9.1*) requires a fine balance in the amounts and activities of the lipoproteins, and relevant enzymes and receptors. The activities of some of these are under hormonal control (e.g. lipoprotein lipase activity is dependent upon the presence of insulin). Imbalances are apparent in clinical conditions such as diabetes mellitus. Genetic disorders have been described which lead to atherosclerosis and coronary heart disease and also to obesity.

9.1.2 Detection

One of the quickest methods that requires only small quantities of material is gradient polyacrylamide gel electrophoresis under nondenaturing conditions (see, for example, refs 1–3). It is usually preferable to apply this method to subfractions of blood plasma derived by density gradient ultracentrifugation. In this case the presence of a protein/lipid band of appropriate electrophoretic mobility derived from a defined part of the density gradient usually suffices to define the class of lipoprotein. Detection on the electrophoretogram of protein is usually by staining with Coomassie blue and of lipid by Oil Red O. Detection by the latter stain is not complicated by the presence of plasma proteins, whereas detection by Coomassie blue may well be.

The presence of lipoproteins in the eluent from columns of sizing gels is best detected by absorption of light at 280 nm. A similar approach to detecting lipoproteins separated by density gradient ultracentrifugation is often used.

9.1.3 Isolation and purification

The standard procedures for separation of lipoprotein classes from each other are based on density gradient centrifugation taking advantage of the differences in density shown in *Table 9.1* (see, for example, refs 1–3). It is usual for the plasma to be derived from fresh blood in the presence of EDTA (1 mg ml^{-1}) to avoid coagulation and also to chelate heavy metal ions which may otherwise increase autoxidation of phospholipid fatty acids as well as inhibiting phospholipases that may attack phospholipids. Addition of an antioxidant, e.g. vitamin E, may be a wise precaution. If the apolipoproteins are to be studied a cocktail of protease inhibitors is also recommended. After removal of cells by centrifugation the plasma may be subjected to sequential flotation ultracentrifugation or to density gradient ultracentrifugation in a swing-out rotor using a stepwise gradient of densities of potassium bromide solution over the range 0.94–1.31 g ml^{-1} (from top to bottom of centrifuge tube) appropriate for the density differences given in *Table 9.1*. Particular attention should be paid to the nature of the centrifuge, rotor, temperature, centrifugation speed and time of the 'run'. For example, a set of parameters adopted by many workers has been to use a Beckman centrifuge with an SW41 rotor, at 20°C and 41 000 rpm for 24 h. The sample in a solution of potassium bromide (1.31 g ml^{-1}) is added to the bottom of the centrifuge tube below the stepwise gradient of potassium bromide solutions as shown in *Figure 9.2*. At the end of the centrifugation layers of lipoprotein can be seen as light-scattering bands at the density interfaces produced by the lipoprotein in question floating on the layer of potassium bromide with the next higher density relative to that of the lipoprotein (see *Figure 9.2*). To obtain reasonably pure preparations of a particular class of lipo-

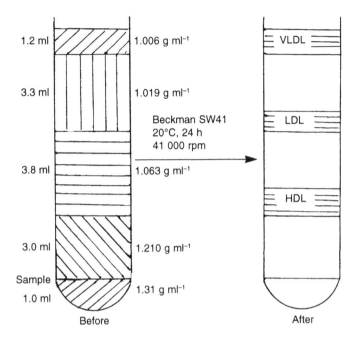

FIGURE 9.2: *Schematic representation of density gradient ultracentrifugation of a sample of human plasma from a fasting individual (i.e. lacking CM). The arrangement of the sample (in KBr, 1.31 g ml⁻¹, 1 ml) and the stepwise density gradient of KBr above it in the centrifugation tube of a swing-out rotor is indicated. The idealized distribution of the main lipoproteins in the gradient after centrifugation is also shown.*

protein a repeat centrifugation will probably be required. The bands of lipoprotein can be removed by careful use of a capillary pipette or by tube puncture.

A convenient and rapid method used for separation of the lipoproteins from each other depends upon precipitation by polyanions. With careful attention to conditions such as temperature and concentrations of mixtures of polyanions and metal ions, precipitated fractions correspond approximately to those separated by density gradient centrifugation. For example, heparin (0.25% w/v) plus Mg^{2+} or Ca^{2+} (0.1 M) will precipitate CM and VLDL but not LDL and HDL whereas heparin (0.1% w/v) plus Mn^{2+} or Co^{2+} (0.05 M) leaves only HDL in solution. In some protocols dextran sulfate (at different concentrations) replaces heparin. Several workers prefer to use sodium phosphotungstate (SPT)–Mg^{2+} mixtures as precipitant primarily because SPT is cheaper and is of the same composition from whatever source. SPT (0.05% w/v)–Mg^{2+} (0.1 M) will precipitate CM and VLDL,

SPT (0.2% w/v)–Mg^{2+} (0.1 M) will precipitate LDL and SPT (2.0% w/v)–Mg^{2+} (0.2 M) will precipitate HDL [3].

Each class of lipoprotein contains several sub-classes. Further purification for separate study of these may be based upon size (e.g. involving rate zonal ultracentrifugation or gel filtration), density (by continuous density gradient ultracentrifugation or zonal ultracentrifugation), charge (by electrophoresis) and type of protein or carbohydrate (by affinity chromatography using an antibody or lectin column). In this way, for example, three different forms of LDL and of HDL have been isolated and characterized. Further details on these procedures should be obtained from standard texts (see e.g. refs 1–3).

Isolation and purification of the lipid components of plasma lipoproteins is essentially as described under the individual lipids in previous chapters.

9.1.4 Quantitation

Due to the complex and different composition of the lipoprotein classes and sub-classes the accurate compositional quantitation of particular lipoproteins is difficult and, probably for most purposes, is not very useful. However, for clinical purposes some analysis is very important because the concentration of HDL-cholesterol in human serum is a negative risk factor for coronary heart disease. Measurement of this parameter (see Section 4.1.5) on samples of serum or plasma after polyanion precipitation of the other lipoproteins is quite common. If the amount of HDL needs to be determined, an immunochemical assay (see Sections 2.4.7, 2.5.4 and 2.5.5) aimed at apolipoproteins A$_1$ of HDL may not be antigenically available and that any one antiserum may not react with all of the isoforms of the apolipoprotein found in a population of patients. Careful polyanion precipitation has enabled subfractionation of HDL into HDL$_2$ (precipitated) and HDL$_3$ with subsequent determination of HDL$_3$-cholesterol and calculation of HDL$_2$-cholesterol by subtraction from total HDL cholesterol. A low HDL$_2$-cholesterol has been correlated with higher risk of early coronary heart disease.

LDL-cholestrol is often calculated rapidly by subtraction of HDL-cholesterol from total plasma cholesterol. The calculation takes into account the ratio of cholesterol:triglyceride in VLDL and LDL and the normal proportion of plasma triglyceride found in VLDL. The calculation requires a triglyceride assay for total plasma (see Section 6.1.3). Alterantively, the LDL-cholesterol is determined directly on LDL isolated by density gradient centrifugation or polyanion precipitation. Immunoassay of apolipoprotein B_{100} also provides an estimate of LDL.

Confirmation of the ratio of the different lipoproteins in a sample of human serum and in the separated fractions may be gained by gradient polyacrylamide gel electrophoresis. Differential staining for protein, lipid and cholesterol wil provide semi quantitative information (see Section 9.1.2).

9.2 Fatty acyl proteins and prenyl proteins

9.2.1 Biological significance

Of the several post-translational modifications of proteins the covalent attachment of lipid-soluble groups is among the more recent discoveries [4]. The proteins so modified include a number of important growth control proteins, cell-surface receptors, cell adhesion molecules, cytoskeletal proteins and viral proteins. The function of the hydrophobic lipid groups on the proteins is not always clear. In some cases they appear to facilitate protein interactions, possibly by direct hydrophobic interactions or by causing a change in protein conformation and the exposure of binding sites. In other cases they appear to have a role in the association, often temporary, of proteins with the lipids of membranes. This is particularly so in the case of some of the evergrowing families of *ras* and related G-proteins, some of which (*rab*) appear to be involved in intracellular vesicle trafficking.

9.2.2 Structures

Fatty acids esterified to proteins include primarily myristic, palmitic and stearic acids. Most studies have concentrated on the first two. Palmitic acid may be present as an oxyester to a serine or threonine residue but more frequently is bound to a cysteine residue through a thioester (*Figure 9.3a*). Palmitoylation of *ras* oncogene proteins requires prior prenylation (see later). Myristic acid is attached to proteins through an amide bond with an N-terminal glycine (*Figure 9.3b*). The acid replaces a methionine residue originally present in the nascent peptide. Some bacteria produce lipoproteins that contain both an *N*-fatty acyl group on an N-terminal cysteine residue of a protein and also a diglyceride residue linked by a thioether bond to the same cysteine residue (*Figure 9.3c*).

Basically two prenyl chains have been found covalently linked to proteins: farnesyl and geranylgeranyl (C_{15} and C_{20}). In both cases the

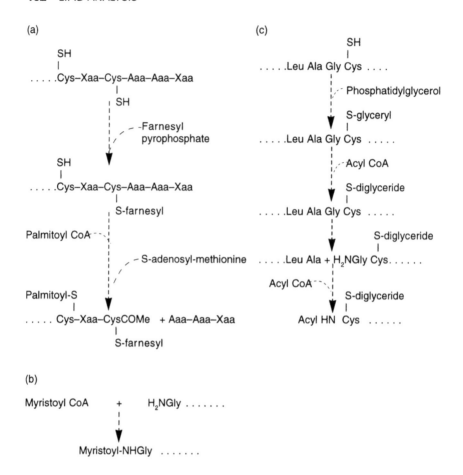

FIGURE 9.3: *Examples of covalent lipid modification of proteins: (a) S-prenylation, S-acylation and O-methylation, where Aaa are aliphatic amino acids and Xaa are any amino acids at the carboxyl-end of a peptide chain; (b) N-myristoylation of an N-terminal glycine of a peptide chain; (c) S-glycerylation (and subsequent acylation to S-diglyceride) and N-acylation of the N-terminal glycine of a peptide chain to produce bacterial lipoprotein.*

link is to a cysteine residue via a thioether bond. The evidence suggests that prenyl pyrophosphate donates the prenyl chain to a cysteine residue of a CAAX C-terminal tetrapeptide box, where A is an aliphatic amino acid and X can be any amino acid (*Figure 9.3a*). There is some indication that if X is methionine, alanine or serine, farnesylation is favored but if X is leucine or phenylalanine, geranylgeranylation occurs. Immediately after prenylation the AAX tripeptide is lost, usually to be replaced by a methyl group. In the case of some *ras* proteins this process is followed by palmitoylation of another cysteine residue further along the peptide chain.

9.2.3 Detection

The presence of an acylated protein in cultures of micro-organisms or of eukaryotic cells can be detected by growing the cells briefly in the presence of a radiolabeled fatty acid and observing the association of radioactivity with the protein throughout its purification. A similar approach may be adopted for prenylated proteins, commencing with radioactive mevalonate as precursor, provided the organism is able to take up this material from the medium.

Observation of the hydrophobic nature of the protein, for example on reversed phase HPLC, may also indicate lipid modification. The availability of antibodies specific to acylated or prenylated proteins also presents a method of their specific detection.

9.2.4 Isolation and purification

The increased hydrophobicity of alkylated and prenylated proteins makes phase separation using Triton X-114 a useful preliminary step in isolation, especially of membrane-bound proteins. The biological system is incubated with Triton X-114 (1%, w/v) at 0–4°C for a few minutes. The extract recovered after refrigerated centrifugation is then warmed to 30–37°C. At this higher temperature two phases are formed. The lower, detergent-rich phase contains the alkylated or prenylated protein.

Myristoylated peptides may then be recovered relatively pure by reversed phase HPLC using propan-1-ol–water mixtures. Palmitoylated prenylated proteins may be too hydrophobic for this system, resulting in difficulty recovering the protein from the column. In this case preparative sodium dodecyl sulfate–polyacrylamide gel electrophoresis (SDS–PAGE) may be the best solution, provided a biologically active product is not required.

O-linked fatty acyl groups may be released from alkylproteins by brief treatment with sodium methoxide (0.5 M) at 100°C. Thioesters are probably best cleaved by hydroxylamine (1 M, 4 h, room temperature). Both O- and N-linked fatty acyl groups are released by treatment with methanolic hydrochloric acid (1 M, 1 h, 100°C). Further isolation and purification of these fatty acids is described in Chapter 5.

Bacterial lipoprotein is usually purified by immunoprecipitation (using a specific antiserum and protein A) followed by preparative SDS–PAGE. If the bacterium was grown on [2-^{3}H]glycerol, [^{35}S]cysteine and/or [^{14}C]palmitate, monitoring of radioactivity during this process and autoradiography of the electrophoretogram produced should be helpful. Mild alkaline (as above) treatment will release the fatty acids from the glyceryl moiety. Acid hydrolysis (methanolysis, as above) will release all fatty acids from the lipoprotein. Acid hydrolysis of performic-acid oxidized lipoprotein releases glyceryl-cysteine sulfone as a separate entity. This can be best identified by high voltage paper electrophoresis against standards. It will clearly retain ^{3}H and ^{35}S if the bacterium was grown in the presence of [^{3}H]glycerol and [^{35}S]cysteine.

Two methods have been used for cleavage of the prenyl S–ether linkage. One of these involves the use of methyl iodide. This produces a methyl sulfonium salt which upon reaction with water gives the prenol as the major product. An alternative method involves the use of Raney nickel which catalyzes the production of the corresponding prenyl hydrocarbon. The prenyl alcohol or prenyl hydrocarbon is probably then best recovered and/or identified by GLC as described in Chapter 4 or 3, respectively.

9.2.5 Quantitation

The lipid-modified proteins are probably best quantified by the application of standard protein methods. This may also require confirmation of the lipid-modified version by chromatographic or electrophoretic methods. HPLC using a UV detector at 280 nm may provide direct identification and quantitation. If prepared from radioactive precursors, radioassay of the purified material may be satisfactory. Alternatively, immunochemical methods (e.g. ELISA or RIA) can be used if antibodies to the protein or preferably to the lipid–protein are available.

Quantitation of released fatty acids, prenols or prenyl hydrocarbons is as described in Chapter 5, 4 or 3, respectively.

References

---

1. Plasma lipoproteins (1986) *Meth. Enzymol.,* **128.**
2. Converse, C.A. and Skinner, E.R. (eds) (1992) *Lipoprotein Analysis.* IRL Press, New York.
3. Mills, G.L., Love, P.A. and Weech, P.K. (1984) in *Laboratory Techniques in Biochemistry and Molecular Biology,* Vol. 14 (R.H. Burdon and P.H. van Knippenberg, eds). Elsevier, Amsterdam.
4. Hooper, N.M. and Turner, A.J. (eds) (1992) *Lipid Modification of Proteins.* IRL Press, New York.

Appendix A

Glossary

Amphipathic: description of a compound containing both hydrophobic (nonpolar, water hating) and hydrophilic (polar, water loving) parts. Such compounds are often soluble in both lipid and aqueous solvents, occur at the interface of mixed lipid/aqueous systems and may act as detergents.

Autoradiography: *see* Fluorography.

Chemical shift (δ)**:** the position of an absorption peak in an NMR spectrum relative to that of tetramethylsilane in units of parts per million, both positions being measured in hertz.

Chromatographic elution: the removal of a solute from a chromatographic column by a solvent (eluant). Isocratic elution is achieved by an eluant of constant composition whereas gradient elution involves a solvent of gradually changing composition. The resulting pattern of solute elution constitutes the elution profile. The volume of solvent required to elute half of a solute (i.e. to the top of the peak on the elution profile) is termed the retention volume (elution volume) of the solute. The time from the commencement of chromatography to elution of half of a solute is termed its retention time (elution time). A good resolution (separation) of the components of a mixture of solutes is achieved if the relative retention volumes (times) of the peaks for the components in the elution profile are high and the peak widths are small.

Chromphore: a chemical group or compound that absorbs light, usually in the visible region.

Eluant: *see* Chromatographic elution.

Elution profile: *see* Chromatographic elution.

Enzyme-linked immunosorbent assay (ELISA): *see* Immunoassay.

Fluorography: detection of the position of radioactivity on a paper or thin layer chromatogram by causing the radiation to stimulate a scintillant (applied to the chromatogram) to produce pulses of light which darken a photographic film placed in close contact with the chromatogram. This is especially useful for detection of tritium which, because of its weak β-radiation, takes a long time to darken a photographic film directly as, for example, used in

165

classical autoradiography without use of a scinitillant for the detection of ^{32}P.

Gradient elution: *see* Chromatographic elution.

Hydrophilic: *see* Amphipathic.

Hydrophobic: *see* Amphipathic.

Immunoassay: a method for quantitation of a compound based on the use of an antibody specific to it. Use of a radioactive sample of pure compound as a standard and of a substoichiometric amount of antibody forms the basis of radioimmunoassay (RIA). The effect of the unlabeled compound on the amount of radioactivity binding to the antibody is measured. Use of a radioactive, nonprecipitating antibody in stoichiometric excess gives rise to immunoradiometric assay (IRMA). In this case, the compound being assayed, plus and minus a standard, is usually first immobilized on a solid surface. If the antibody is labeled, not with radioisotope but by covalent addition of an enzyme, the method is called an enzyme-linked immunosorbent assay (ELISA).

Immunoradiometric assay (IRMA): *see* Immunoassay.

Isocratic elution: *see* Chromatographic elution.

Radioimmunoassay (RIA): *see* Immunoassay.

Resolution: *see* Chromatographic elution.

Retention (elution) time: *see* Chromatographic elution.

Retention (elution) volume: *see* Chromatographic elution.

Safety hazard: the potential of a compound or system to endanger health.

Safety risk: the possibility of a safety hazard being realized taking account of the procedures and intensity or amount of hazardous compound or system being used.

SI units: an internationally agreed system of units of measurement of physical and chemical parameters.

Appendix B

Suppliers

Alltech Assoc., 6–7 Kellet Road Industrial Estate, Camforth, Lancs LA5 9XP, UK. Tel. (0)1524 734451; Fax (0)1524 733599.
Alltech Assoc. Inc., 2051 Waukegan Road, Deerfield, IL 60015, USA. Tel. 708 948 8600; Fax 708 948 1078.

Amersham International plc, Amersham Place, Little Chalfont, Bucks HP7 9MA, UK. Tel. 0800 616929; Fax 0800 616927.

Anachem Ltd, Anachem House, 20 Charles Street, Luton, Beds LU2 0EB, UK. Tel. (0)1582 456666; Fax (0)1582 391768.

Analtech Inc., 75 Blue Hen Drive, PO Box 7558, Newark, DE 19711, USA. Tel. 302 737 6960; Fax 307 737 7115.

ATI Unicam, York Street, Cambridge CB1 2PX, UK. Tel. (0)223 358866; Fax (0)223 312764.

BDH Ltd, Hunter Boulevard, Magna Park, Lutterworth LE17 4XN, UK. Tel. 0800 223344; Fax (0)1455 558586.

Beckman Instruments UK Ltd, Oakley Court, Kingsmead Bus Park, London Road, High Wycombe, Bucks HP11 1JU, UK. Tel. (0)1494 441181; Fax (0)1494 447558.
Beckman Nuclear Systems Operations, 2500 Harbor Boulevard, Fullerton, CA 92634, USA. Tel. 714 871 4848/800 742 2345; Fax 800 643 4366.

Berthold Instruments (UK) Ltd, 20 Vincent Avenue, Crownhill Business Centre, Milton Keynes, Bucks MK8 0AB, UK. Tel. (0)1908 265744; Fax (0)1908 265956.

Bio-Rad Laboratories Ltd, Bio-Rad House, Maylands Avenue, Hemel Hempstead, Herts HP2 7TD, UK. Tel. 0800 181134; Fax (0)1442 259118.

Boehringer Mannheim GmbH, Biochemica, PO Box 31 01 20, D-6800 Mannheim 31, Germany. Tel. 6221 7591; Fax 6221 759 8509.
Boehringer Mannheim GmbH (UK), Boehringer Corporation (London) Ltd, Bell Lane, Lewes, Sussex BN7 1LG, UK. Tel. (0)1273 480444.

Camlab Ltd, Nuffield Road, Cambridge CB4 1TH, UK. Tel. (0)1223 424222; Fax (0)1223 420856.

Canberra-Packard Ltd, Brook House, 14 Station Rd, Pangbourne, Bucks RG8 7DT, UK. Tel. (0)1734 844981; Fax (0)1734 844059.
Packard Instrument Co., 2200 Warrenville Road, Downers Grove, IL 60515, USA. Tel. 708 969 6000; Fax 708 969 6511.

Chromacol Ltd, Glen Ross House, 3 Little Mundell, Mundell Industrial Centre, Welwyn Garden City, Herts AL7 1EW, UK. Tel. (0)181 368 7666; Fax (0)181 361 4698.

Chrompack UK Ltd, Unit 4, Indescon Court, Millharbour, London E14 9TN, UK. Tel. (0)171 515 8080; Fax (0)171 538 0908.

Dionex (UK) Ltd, 4 Albany Court, Camberley, Surrey GU15 2PL, UK. Tel. (0)1276 691722; Fax (0)1276 691837.

Dyson Instruments Ltd, Hetton Lyons Industrial Estate, Hettong, Houton-le-Spring, Tyne and Wear DH5 0RH, UK. Tel. (0)1783 260452; Fax (0)191 5170844.

Fisons plc Scientific Division, Bishop Meadow Road, Loughborough, Leics LE11 0RG, UK. Tel. (0)1509 231166; Fax (0)1509 231893.

Fluka Chemie AG, Industriestrasse 25, CH 9470, Buchs, Switzerland. Tel. 81 755 2511; Fax 81 756 5449.

Hewlett Packard Ltd, Heathside Park Road, Cheadle Heath, Cheshire SK3 0RB, UK. Tel. (0)1345 125292; Fax (0)161 495 5531.

Hichrom Ltd, 1 The Markham Centre, Station Road, Theale, Reading, Berks RG7 4PE, UK. Tel. (0)1734 303660; Fax (0)1734 323484.

V.A. Howe Ltd, 12–14 St Ann's Crescent, London SW18 2LS, UK. Tel. (0)1295 252666; Fax (0)1295 268096.

HPLC Technology Ltd, Wellington House, 10 Waterloo Street, West Macclesfield, Cheshire SK11 6PJ, UK. Tel. (0)1625 613848; Fax (0)1625 616916.

JEOL UK Ltd, Jeol House, Silver Court, Watchmead, Welwyn Garden City, Herts AL7 1LT, UK. Tel. (0)1707 377117; Fax (0)1707 373254.

Jones Chromatography Ltd, Tir-y-Berth Industrial Estate, New Road, Hengoed, Mid Glamorgan CF8 8AU, UK. Tel. (0)1443 816991; Fax (0)1443 816552.

Kodak Clinical Diaignostics Ltd, Mandeville House, 62 The Broadway, Amersham, Bucks, UK. Tel. (0)1494 431717; Fax (0)1494 725301.

Kontron Instruments Ltd, Blackmoor Lane, Crosley Business Park, Watford, Herts WD1 8XQ, UK. Tel. (0)1923 245991; Fax (0)1923 922012.

Macherey-Nagel & Co., D-5160 Duren, Werkstrasse 6-8, Germany. Tel. 2421 9690; Fax 2421 62054.

Merck Ltd, Merck House, Selsdown Lane, Poole, Dorset BH15 1TD, UK. Tel. (0)1202 669700; Fax (0)1202 666536.
E. Merck, Frankfurterstrasse 250, D6100, Darmstadt 1, Germany. Tel. 615 17200; Fax 615 17222.

New England Nuclear, Du Pont (UK) Ltd, Medical Products Dept, Biotechnology Systems Division, Wedgwood Way, Stevenage, Herts SG1 6YH, UK. Tel. (0)1438 734026; Fax (0)1438 734379.

Perkin Elmer Ltd, Post Office Lane, Beaconsfield, Bucks HP9 1QA, UK. Tel. (0)1494 274411; Fax (0)1494 679333.

Pharmacia Biotech Ltd, 23 Grosvenor Road, St Albans, Herts AL1 3AW, UK. Tel. (0)1727 214000; Fax (0)1727 814020.
Pharmacia LKB Biotechnology AB, Bjorkgatan 30, S-751 82 Uppsala, Sweden. Tel. 18 165000; Fax 18 143820.
Pharmacia LKB Biotechnology, 800 Centennial Ave, PO Box 1327, Piscataway, NJ 08855-1327, USA. Tel. 201 457 8000; Fax 201 457 0557.

Phase Separations Ltd, Deeside Industrial Park, Deeside, Clwyd CH5 2NU, UK. Tel (0)1244 289289; Fax (0)1244 289500.

Severn Analytical Ltd, 10 Waterloo Street West, Macclesfield, Cheshire SK11 6PJ, UK. Tel. (0)1625 613848; Fax (0)1625 616916.

Shimadzu Corporation, International Marketing Division, Shinjuku Mitsui Building, 1-1, Nighi-Shinjuku 2-chome, Shinjuku-ku, Tokyo 163, Japan. Tel. (0)3 219 5641; Fax (0)3 219 5710.

Sigma Chemical Co.. Ltd, Fancy Road, Poole, Dorset BH17 7TG, UK. Tel. (0)1202 733114; Fax (0)1202 715460.
Sigma Chemical Co., PO Box 14508, St Louis, MO 63178, USA. Tel. 800 2487791.

Supelco UK, Fancy Road, Poole, Dorset BH17 7NH, UK. Tel. 0800 887733; Fax 0800 378785.

Varian Assoc. Ltd, 28 Manor Road, Walton on Thames, Surrey KT12 2QF, UK. Tel. (0)1932 243741; Fax (0)1932 228769.

Waters Corporation, 34 Maple St, Milford, MA 01757-3696, USA. Tel. 508 478 2000; Fax 508 482 3361.

Whatman International Ltd, St Leonard's Road 20/20, Maidstone, Kent ME16 0LS, UK. Tel. (0)1622 676670; Fax (0)1662 677011.

Appendix C

Further reading

Most general references relevant to specific topics are given at the end of each chapter. Those listed below provide further background material. Some very general texts are listed here as well as at the end of specific chapters.

Christie, W.W. (1987) *High Performance Liquid Chromatography and Lipids*. Pergamon Press, Oxford.

Christie, W.W. (1989) *Gas Chromatography and Lipids*. The Oily Press, Ayr, UK.

Gunstone, F.D., Harwood, J.F. and Padley, F.B. (eds) (1994) *The Lipid Handbook* (2nd edn). Chapman and Hall, London.

Gurr, M.I. and Harwood, J.L. (1991) *Lipid Biochemistry, an Introduction* (4th edn). Chapman and Hall, London.

Hamilton, R.J. and Hamilton, S. (eds) (1992) *Lipid Analysis: A Practical Approach*. IRL Press at Oxford University Press, Oxford.

Harwood, J.L. and Russell, N.J. (1984) *Lipids in Plants and Microbes.* George Allen and Unwin, Hemel Hempstead.

Kates, M. (1985) *Techniques of Lipidology* (2nd edn). Elsevier, Amsterdam.

Mead, J.F., Alfin-Slater, R.B., Hawton, D.R. and Popjak, G. (1986) *Lipids: Chemistry, Biochemistry and Nutrition*. Plenum Press, New York.

Routledge, C. and Wilkinson, S.G. (eds) (1988) *Microbiol Lipids,* Vol. 1. Academic Press, London.

Vance, D.E. and Vance, J.E. (eds) (1985) *Biochemistry of Lipids and Membranes.* Benjamin/Cummings, Manloe Park, CA, USA.

Index

To be used in conjunction with the detailed list of contents (pp. v-ix).

Absorbance, light, 33
Absorption of light
  detectors of, 9, 18
Acetylcholinesterase
  and PI glycan, 123
Alanine
  in bacterial lipids, 144, 150
Anionic exchangers, 30, 31
Antibodies, 49–57
  in ELISA, 56, 57
  in hemagglutination, 52–54
  in immunostaining, 54, 55
  in IRMA, 55, 56
  in RIA, 49-51, 56
  to lipids, 51, 52
Antioxidant
  protection, 6, 76, 158
  tocopherols as, 80, 154, 158
  xanthophylls as, 80
Apolipoproteins, 154–157, 160
Arachidic acid, 102
Arachidonic acid, 103–105
Atherosclerosis, 157

Bactoprenols, 83, 142
Band broadening
  in chromatography, 12
Batyl alcohol, 110
Becquerel, 47
Betulaprenols, 83
Bials reagent, 133, 139
Bligh and Dyer
  extraction, 5

Calibration, 3
Carcinogens, 62
Cardiolipin, 116
Carotenes, 73–78
  (see also Xanthophylls)
Carr–Price reaction, 89, 91, 95
Cationic exchanger, 31

Cephalin, 113
Ceramide, 113, 130
  glycanase, 138
Cerebronic acid, 101
Cerebrosides, 101, 115
Cerenkov counting, 38, 45
Chaulmoogric acid, 102
Chimyl alcohol, 110
Cholecalciferol, 80, 84, 89, 91
Cholesterol, 54, 55, 79, 84, 87, 89–92, 102
  esterase, 92
  esters 90, 92, 102, 135
  lipoproteins and, 153–157, 160
  oxidase, 92
Chromanols, 80, 85
Chromatography
  borate, 32
  conventions, 13, 14
  detectors, 9, 16, 18, 19, 20, 22, 24, 25
  GLC, 9, 18, 22–24
  HPAEC, 31, 141
  HPLC, 15–20, 30–32, 76–78
  immunostaining, 54, 55
  ion-pair, 32
  of fatty acids, 58–60, 102–105
  of monosaccharides, 141
  Sep-Pak® and, 20, 21, 76, 90, 136
  silver nitrate, 32, 76, 103, 108, 109
  TLC, 9, 10, 26-30, 76, 78, 103, 108, 109
Chromophoric groups, 33
Citronellol, 82
COSHH regulations, 62, 67
Cutin, 110
Cyclooxygenase, 105

Dolichols, 79, 83, 89–90
  in N-glycosylation, 142, 143, 145, 146
Dragendorff reagent, 116

Ehrlich's reagent, 133, 139
ELAM-1, 129
Electron capture detector, 24, 25
ELISA, 56, 57
Elson–Morgan test, 133
Eluants, polarity, 7, 8
Elution volume, 13
Emmerie–Engel reaction, 91, 99
Enzyme immunoassay, 56
EPA, 103
Ergocalciferol, 80, 84, 89, 91
Ergosterol, 79, 84, 90
Errors in analysis, 2
E-selectin, 129
Ethanolamine in PI glycan, 123

Farnesal, 94
Farnesane, 73, 74
Farnesol, 82, 161, 162
Fatty acyl sugars, 147
Ficaprenols, 83
Flame ionization detectors, 18, 24, 25
Fluorescence detectors, 9, 18
Fluorography, 39, 40
Folch extraction, 5, 134, 148
Fucose, 131
Fucosidase, 137

Galactose
    in glycosphingolipids, 131, 132,
        134, 140, 141
    in glycosylglycerides, 146–148
    in polyisoprenoid derivatives, 144
Galactosyl ceramide, 130
Geiger–Müller tube, 45, 46
Geranial, 94
Geraniol, 82
Geranylgeranial, 94
Geranylgeraniol, 82, 161, 162
Glucosamine in PI glycans, 123
Glucose
    in glycosphingolipids, 131, 140,
        141
    in glycosylglycerides, 147
    in phosphoglycolipids, 150, 151
    in PI glycans, 123
    in polyisoprenoid derivatives, 143,
        145
Glucosyl ceramide, 130
Glycolipids of lipoproteins, 153
Glycoprotein
    in lipoproteins, 153
    N-glycosylation, 142, 143
    O-mannosylation, 143

Hemagglutination, 52, 53

Immunoradiometric assay, 55, 56
Immunostaining, 54, 55
Inositol phosphates, 118–122

Kojibiose, 150

Lactobacillic acid, 102
Lambert–Beer Law, 33
Lauric acid, 102
Lecithin, 113, 157
Lectins
    in lipoprotein purification, 160
    TLC stain, 54, 134
Liebermann–Burchard
    reaction, 88, 91, 148
Lignoceric acid, 101
Linoleic acid, 101–104
Linolenic acid, 103, 147
Lipases in TAG analysis, 59, 60
Lipoprotein, bacterial, 163
    lipase, 155–157
    receptors, 155, 156
Liposomes, 54, 55, 61
Lipoxygenase, 105
Liquid scintillation
    counting, 38, 41–44

Mannose
    in glycolipids, 140, 141
    in glycosylglycerides, 147
    in PI glycans, 123
    in polyisoprenoid derivatives,
        143–145
Mass transfer, 12
Menaquinones, 95–97, 99
Methanolysis, 6
Mobile phases, 7, 10, 12, 15
    in GLC, 22
    in HPLC, 17
    in TLC, 26–29
Monosaccharides
    of glycosphingolipids, 130–132,
        137, 138, 141
Morgan–Elson test, 133, 134
Myristic acid, 102, 161–163

N-acetylgalactosamine
    in glycosphingolipids, 131–134,
        137
N-acetylglucosamine
    in glycosphingolipids, 131, 133
    in polyisoprenoid derivatives,
        143–145

N-acetylhexosaminidase, 137
N-acetylmuramic acid, 144
N-acetylneuraminic acid
    in glycosphingolipids, 131–135,
        137
Nerol, 82
Nerolidol, 82
Neuraminidase, 137
Ninhydrin reagent, 116
N-myristoylation, 161, 162

Obesity, 157
Oils, plant, TAGs of, 109
Oleic acid, 103, 104
Optical density, 33

Palmitic acid, 102, 104, 161, 162
Palmitoleic acid, 103
Partition coefficient, 12
Pathogenic organisms, 67
Periodic acid-Schiff reaction, 133
Pheromones, 79, 93
Phosphatidyl-
    choline, 54, 55, 61, 113,
        114, 116, 117, 157
    ethanolamine, 113, 114, 116, 117
    glycerol, 116
    inositol, 116–120
    serine, 113, 114, 116, 117
Phospholipase
    A_1, A_2 and D, 57–59
    C, 57–59, 119–125
    inhibition, 6
    PI specific C, 122
Photomultiplier tube, 41, 42, 44, 45
Phylloquinone, 95–97
Phytadienes, 74
Phytane, 73, 74
Phytanic acid, 101, 102
Phytanol and derivatives, 79–81, 89
Phytic acid, 119
Phytol, 81, 82, 87, 89
Plasmalogens, 92–93, 94, 110, 111,
        114, 115, 124
Plastoquinone, 95–99
Polarity, chemical, 7, 8
Polyprenols, 79, 82, 83, 87, 89, 142,
        144, 145
Presentation of results, 3
Pristane, 73, 74
Pristanol, 79–81
Proportional counting, 45
Prostacyclin, 105, 106
Prostanoic acid, 105
Prostanoids, 105, 106

Protein
    A, 54, 55
    N-myristoylation, 161, 162
    O-palmitoylation, 161
    S-glycerylation, 162
    S-palmitoylation, 161, 162
    S-prenylation, 162

Quenching, 42–44

Radioactivity detectors (flow
        through), 44
Radioisotopes
    becquerel, Curie, 47
    β-radiation of, 38–41, 43, 45, 46
    $E_{\beta max}$, 38
    γ-radiation of, 39, 43, 46, 54
    in lipid analysis, 38
    RIA and, 48, 49
    safety of, 63, 64, 66, 67
    specific radioactivity, 47, 48
Refractive index, 9, 19
Retention time/volume, 12, 13
Retinal, 93–95
Retinol, 73, 80, 82, 83, 87, 89, 91
R_f value, 28
Rhamnose
    in glycosylglycerides, 147
    in polyisoprenyl derivatives, 144
RIA, 49, 50
Ricinoleic acid, 101
Risk assessment, 62, 63

Safety
    carcinogens and, 62
    chemicals disposal, 65, 67
    chemical hazard
        definitions, 64, 65
    clothing, 69
    controlled substances, 62
    COSHH and, 62, 67
    electricity and, 69
    gas cylinders and, 68
    genetic change and, 68
    protection, 69, 70
    sharp instruments and, 68
    storage and, 63
Schiff's reagent, 93–94, 133
Scintillants, 39, 41, 42, 44
Selachyl alcohol, 110
Sialic acid – see N-acetylneuraminic
        acid
SIDA, 49
Sitosterol, 79, 84
Solanesol, 83

Spadicol, 83
Sphingomyelin, 114, 115
Sphingosine, 85, 86, 89, 91, 92
  in glycosphingolipids, 129, 130,
    132, 139
  in sphingomyelin, 113, 117
Squalene, 73, 74
Stationary phase
  general, 7, 9, 10, 12
  GLC and, 22, 23, 24, 76, 103, 104
  HPLC and, 17, 77
  TLC and, 26
Stearic acid, 102, 104
Sterculic acid, 102
Stereoisomerism, 81–83
Steroid hormones, 79, 85
Steroids, 80, 84, 87
Sterol glycosides, 84, 147
Stigmasterol, 79, 84
Suberin, 110
Sulfoquinovose, 126, 147, 148
Svennerholm
  and ganglioside isolation, 135

Theoretical plates, 14
Thermal conductivity detector, 24, 25
Thromboxanes, 104, 105
Tocopherolquinones, 96, 97, 99, 100
Tocopherols, 80, 85, 87, 89, 91, 99,
    155, 158
Triacylglycerol
  in lipoproteins, 153–157, 160
Trioleylglucose, 147

Ubiquinol, 90, 95, 155
Ubiquinones, 90, 95–100

Vitamin A – *see* Retinol
Vitamin D – *see* Chole- and
    Ergocalciferol
Vitamin E – *see* Tocopherols
Vitamin K – *see* Mena- and
    Phylloquinone

Xanthophylls, 80, 85, 88, 89, 91